Human Growth, Physical Fitness and Nutrition

Medicine and Sport Science

Vol. 31

Series Editors
M. Hebbelinck, Brussels
R.J. Shephard, Toronto, Ont.

Founder and Editor from 1969 to 1984
E. Jokl, Lexington, Ky.

Basel · München · Paris · London · New York · New Delhi · Bangkok · Singapore · Tokyo · Sydney

Human Growth, Physical Fitness and Nutrition

Volume Editors
R.J. Shephard, Toronto, Ont.
J. Pařízková, Prague

44 figures and 69 tables, 1991

KARGER

Basel · München · Paris · London · NewYork · New Delhi · Bangkok · Singapore · Tokyo · Sydney

Medicine and Sport Science

Published on behalf of the International Council of Sport Science and Physical Education

---

Library of Congress Cataloging-in-Publication Data
  Human growth, physical fitness, and nutrition / volume editors, R.J. Shephard, J. Pařízková.
  (Medicine and sport science; vol. 31)
  Includes bibliographical references and index.
  1. Children – Developing countries – Nutrition. 2. Malnutrition – Developing countries.
  3. Physical fitness. 4. Children – Growth.
  I. Shephard, Roy J. II. Pařízková, Jana. III. Series.
  [DNLM: 1. Growth. 2. Nutrition. 3. Physical Fitness.]
  ISBN 3–8055–5275–0

Drug Dosage
  The authors and the publisher have exerted every effort to ensure that drug selection and dosage set forth in this text are in accord with current recommendations and practice at the time of publication. However, in view of ongoing research, changes in government regulations, and the constant flow of information relating to drug therapy and drug reactions, the reader is urged to check the package insert for each drug for any change in indications and dosage and for added warnings and precautions. This is particularly important when the recommended agent is a new and/or infrequently employed drug.

All rights reserved.
  No part of this publication may be translated into other languages, reproduced or utilized in any form or by any means, electronic or mechanical, including photocopying, recording, microcopying, or by any information storage and retrieval system, without permission in writing from the publisher.

© Copyright 1991 by S. Karger AG, P.O. Box, CH–4009 Basel (Switzerland)
  Printed in Switzerland on acid-free paper by Thür AG Offsetdruck, Pratteln
  ISBN 3–8055–5275–0

Contents

Preface VII

Pařízková, J. (Prague): Human Growth, Physical Fitness and Nutrition under Various Environmental Conditions 1
Chen, J.D. (Beijing): Growth, Exercise, Nutrition and Fitness in China 19
Satyanarayana, K.; Prasanna Krishna, T. (Hyderabad); Banerji, D. (New Delhi); Narasinga Rao, B.S. (Hyderabad): Social Epidemiology of Nutrition in the Ranga Reddy District of India and Its Implications for Human Resources Development 33
Spurr, G.B. (Milwaukee, Wisc.); Barac-Nieto, M.; Reina, J.C. (Cali): Growth, Maturation, Body Composition and Maximal Aerobic Power of Nutritionally Normal and Marginally Malnourished School-Aged Colombian Children 41
Dekkar, N. (Algiers): Growth, Nutrition and Physical Performance in Algeria 61
Narváez Pérez, G.E.; D'Angelo, C.P.; Zabala, R.D. (Buenos Aires): Physical Fitness in Children and Adolescents from Differing Socioeconomic Strata 80
Arbesú, N. (Havana): Aerobic and Anaerobic Physical Capacity of Cuban Schoolchildren Subjected to Different Motor Regimens 99
Rocha Ferreira, M.B. (São Paulo); Malina, R.M. (Austin, Tex.); Rocha, L.L. (São Paulo): Anthropometric, Functional and Psychological Characteristics of Eight-Year-Old Brazilian Children from Low Socioeconomic Status 109
Malina, R.M. (Austin, Tex.); Little, B.B.; Buschang, P.H. (Dallas, Tex.): Estimated Body Composition and Strength of Chronically Mild-to-Moderately Undernourished Rural Boys in Southern Mexico 119
Shephard, R.J. (Toronto, Ont.): Somatic Growth and Physical Performance in Canada 133
Pařízková, J.; Heller, J. (Prague): Relationship of Dietary Intake to Work Output and Physical Performance in Czechoslovak Adolescents Adapted to Various Work Loads 156
Shephard, R.J. (Toronto, Ont.): Conclusions 168

Subject Index 171

Preface

The activities of the International Pediatric Work Physiology group over the last 20 years have led to the emergence of a substantial literature on the growth of physical performance in normal healthy children. However, because of resource constraints that limit both data collection and the preparation of technical reports in the third world, the majority of information currently available concerns that small segment of the world fortunate enough to live in Western Europe or North America.

The unfortunate reality of the coming century is that most children will live in desperately poor circumstances in the third world; moreover, their growth, health and physical performance will likely be impaired by an inadequate intake of energy, proteins and vitamins. Nutritionists have reviewed the needs of such populations, but have tended to approach both energy and nutrient requirements from the perspective of a fixed energy demand. In the present volume, the question is addressed from the perspective of the active child, whether the activity arises in the course of child labour, special programmes of physical education, or training in elite sport camps. Several of the contributors write passionately about the privileges and social barriers that enchain not only the current generation of third world children, but (through the likelihood that they will show poor physical and intellectual development as adults) put in jeopardy the lives of future generations from the under-privileged nations. Strong exception is also taken to those from the 'first world' who sustain such social injustice by hypotheses that children can adapt to malnutrition, becoming 'small but healthy'.

Some of the contributors to this monograph will be unfamiliar to pediatric work physiologists; their writing is based on first-hand experiences in the third world. However, in order to provide a perspective on development in more privileged societies, there are also reports from Czechoslovakia, the USA and Canada. The latter two chapters, incidentally, demonstrate that socio-economic gradients have markedly affected child growth in North America until quite recently.

The present text will appeal not only to the specialized group of pediatric work physiologists, but to all of those concerned with the health, nutrition and physical education of the growing child. It may require a major political upheaval to remedy the social injustices documented in this volume, but if the authors have done no more than indicate the need to distribute world resources in a more equitable fashion, they have performed an important service.

Roy J. Shephard, Toronto, Ont., 1990
Jana Pařízková, Prague, 1990

Human Growth, Physical Fitness and Nutrition under Various Environmental Conditions

J. Pařízková

Research Institute for Physical Education, Charles University, Prague, Czechoslovakia

Introduction

The health and fitness of children and youth has become a primary interest in countries throughout the world; the health of children undoubtedly provides the key to the health of the adult population, and indeed to the whole future of our planet. This concern applies especially in developing countries, where growing subjects account for a high proportion of the total population (fig. 1).

About one-third of the total world population are currently under 15 years of age; this figure varies widely according to the region under examination. The highest proportion of young individuals is found in Africa (45%) and Latin America (about 39%). This situation is similar in South Asia; in Eastern Asia, the proportion of the young is somewhat lower, figures of about 30% being found in China and Japan. On the other hand, in Western Europe the proportion of children and youth has decreased to about 18–24%. North American countries are very close to European numbers in this respect, but in the latter countries child morbidity and mortality are markedly lower, with a greater life expectancy at birth, and a longer average survival.

Given the regional differences in the proportion of individuals, their morbidity, mortality and life expectancy, the question arises as to how far such variation can be related to environmental conditions – social, economic, geographic, hygienic, cultural and so forth. The most urgent problem of the developing countries is the high mortality, especially in the first few months of life. Nevertheless, even if children in these countries survive the neonatal period, many further problems arise from continuing adverse life conditions (malnutrition, lack of hygiene, poor health care, inadequate water supplies, child labour, and so forth). All of these chal-

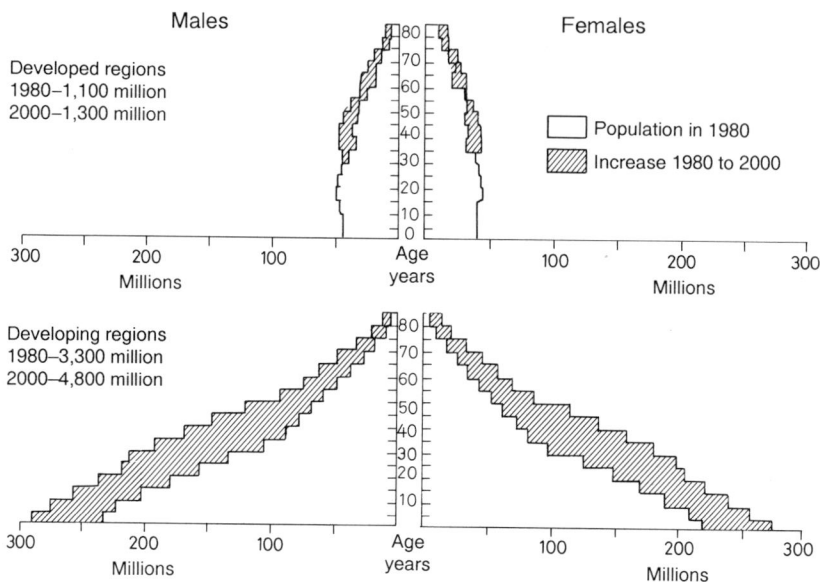

Fig. 1. World population by age and sex, 1980 and 2000. United Nations estimates and projections as assessed by 1980.

lenges not only threaten survival in the following years of childhood, but also limit the achievement and maintenance of developmental potential with respect to mental and physical fitness, along with health status both currently and into adult life and old age.

On the other hand, in parts of the world where better living conditions allow a greater life expectancy, other community problems appear, as the proportion of adult and older subjects becomes higher. Such problems are perhaps less urgent; nevertheless, the 'diseases of civilization' already have an alarming impact upon morbidity and mortality in the industrially developed countries. Ischemic heart disease, atherosclerosis and other metabolic diseases are – at least in part – diseases of affluence. Overabundant and poorly balanced nutrition which exceeds daily energy needs, along with hypokinesia and mental stress starting at an early age, seem to be important causes of this situation. The immediate practical consequences are poor productivity, increased medical care expenditures and a shortened life expectancy.

The definition of appropriate standards of human growth and physical performance under various environmental conditions has to date been incomplete. Growth curves have generally been based on the absence of actual pathological phenomena, with other variables being interpreted relative to body size. These criteria seem insufficient in the context of health, overall fitness and future prognosis. Traditionally, it has been assumed that a child who achieves a certain height and body mass by a particular age also has a satisfactory development with respect to the above-mentioned characteristics.

There is already evidence that even given a proper weight for age, or weight for height, the health status and fitness level of a child may be inadequate. On the other hand, the reverse can also be true, some smaller children being healthier and fitter than their bigger peers.

Recommended dietary allowances (RDA) for children also become problematic when varying life conditions are taken into account. The balancing of energy intake against energy output surely has a key position among the environmental factors needed to safeguard the optimal development of the growing human organism. Keeping in mind the above-mentioned problems, the definition of an "optimal minimum" RDA based on adequate and reliable criteria seems indispensable. Note particularly that the RDA can vary according to the criteria chosen – for example, attainment of a certain body size at a certain age, achievement of a certain level of physical and overall fitness, or maximization of survival. The ideal RDA would consider all of these aspects of development. Sometimes, the several objectives conflict with each other. For example, achievement of a greater body mass may be in conflict with overall health status and survival, as shown both in experimental models with laboratory animals (fed ad libitum or a restricted diet) and in humans (those who survive the longest have often had a very poor level of nutrition in childhood).

There is no doubt that there exist certain bounds of dietary intake, body size and growth beyond which a normal existence is not possible. Nevertheless, there is evidence that when considering overall health and physical performance, the range of dietary intake and body sizes can be much larger than has been generally agreed up till now. In particular, a satisfactory existence can be assured with much lower levels of energy intake than has previously been recommended. The acceptable range varies interindividually according to hereditary factors and adaptive processes affecting both energy input and output, that is, dietary intake and physical activity.

Selected cross-sectional and longitudinal studies of children and adolescents in various parts of the world have already given some basis for an analysis of these problems, but they underscore the necessity to define more exactly the optimal conditions in their full variety with respect to diet and physical work rates during growth and development. An effective recommendation for health care and preventive medicine thus requires the elaboration of precise dietary and work-load standards applicable throughout the life span, with special attention to the needs of childhood and adolescence. According to the evidence gained to date, such recommendations ought also to cover the fetal period, providing standards for the health, nutrition and work-loads sustained by future mothers.

It is usually considered that once a certain level of body mass and physical activity have been established and an appropriate growth rate defined, there can be only one level of food intake at which energy balance can be achieved. As mentioned above, there are quite wide ranges of weight, fatness and energy intake compatible with full health, a high level of physical fitness and full mental fitness. Moreover, a real steady state is not usually achieved. A certain imbalance between the input and the output of energy (within the adaptative limits of the organism) may even provide a positive stimulus to achieving the full potential of the human organism.

An appropriate knowledge and analysis of these phenomena is helpful in handling the unfavourable situations which appear in contrasting ways in developing and industrially developed countries. The extremes of energy balance resulting from a reduced energy intake and an excessive energy output where young children perform heavy physical labour, and/or the increased food intake and hypokinesia of the modern city dweller seem highly unfavourable for development of the human organism. As regards the latter lifestyle, there is a further problem that the way of life encountered in the industrially developed countries has been viewed as a desirable model for the developing countries. Thus, very similar health and fitness problems have already appeared in the affluent social strata of developing countries.

Insufficient and conflicting evidence on these issues has resulted from poorly coordinated methodological approaches and unclear definitions and criteria. The establishment of optimal and limiting conditions of energy input and output needs a further analysis of existing knowledge. Even if it seems a remote goal, the achievement of "positive health" (that is the status of full physical, mental and social well-being) ought to be the focus of our interest, since it offers one of the most important opportunities for the amelioration of life conditions and the enhanced economic productivity of all. The population that is fitter and healthier can surely cope better with the most urgent problems of our present world.

Approaches to Fitness Assessment

General health includes not only the absence of disease or infirmity, and/or the achievement of a certain body size, but it also implies realization of an adequate physical fitness and nutritional status.

Morphological Development

Body size and composition have been most often used as characteristics to explore the normality of development during growth and maturation; changes have been related to both calendar and biological age. Sexual differences, too, are apparent from earliest childhood. Environmental factors, including nutrition and physical work-load, can further modify the above-mentioned characteristics in either a positive or a negative way.

Most authors have suggested that the growth potential of children should be achieved relative to such criteria. However, the normality of many other characteristics of the human organism has had little stress till now. Specifically, the functional potential as regards physical and mental fitness deserves greater mention.

Standards of height and body mass have been established and agreed [for example, in the document on 'Energy and protein requirement', Joint FAO/WHO/UNU Expert Consultation in Rome 1981, Tech. Rep. Ser. 724; Geneva, WHO, 1985]. Anthropometric data for children and adolescents (the mass for age of boys and girls from birth up to 10 years) are given in table 1. The median body mass for age and height of boys and girls from 10 to 18 years are given in tables 2 and 3, respectively [WHO, 1985]. These standards are generally recommended for the evaluation of body size at various ages. There are also a number of growth grids and tables, based

Table 1. Anthropometric data of children and adolescents [WHO, 1985]: mass (kg) for age of children

Age years	Boys			Girls		
	−2 SD	median	+2 SD	−2 SD	median	+2 SD
0	2.4	3.3	4.3	2.2	3.2	4.0
0.25	4.1	6.0	7.7	3.9	5.4	7.0
0.5	5.9	7.8	9.8	5.5	7.2	9.0
0.75	7.2	9.2	11.3	6.6	8.6	10.5
1.0	8.1	10.2	12.4	7.4	9.5	11.6
1.5	9.1	11.5	13.9	8.5	10.8	13.1
2.0	9.9	12.6	15.2	9.4	11.9	14.5
3	11.4	14.6	18.3	11.2	14.1	18.0
4	12.9	16.7	20.8	12.6	16.0	20.7
5	14.4	18.7	23.5	13.8	17.7	23.2
6	16.0	20.7	26.6	15.0	19.5	26.2
7	17.6	22.9	30.2	16.3	21.8	30.2
8	19.1	25.3	34.6	17.9	24.8	35.6
9	20.5	28.1	39.9	19.7	28.5	42.1
10	22.1	31.4	46.0	21.9	32.5	49.2

Table 2. Median mass (kg) for age and height of adolescent boys [WHO, 1985]

Height cm	Age, years								
	10	11	12	13	14	15	16	17	18
120									
125	24.2								
130	26.8	27.0							
135	29.3	29.4	29.6						
140	32.2	32.2	32.4	32.4					
145	34.9	35.7	35.4	35.8	36.3				
150	38.1	38.5	39.0	39.1	39.3	39.2			
155		41.5	42.1	42.7	43.4	43.5	44.8		
160			46.2	46.7	47.4	48.0	49.8	51.5	53.9
165				50.9	51.4	52.3	53.1	55.1	57.1
170					55.6	56.5	58.1	59.1	60.5
175					59.7	60.4	61.9	63.5	64.7
180						65.1	65.7	66.1	67.1
185							69.5	70.3	71.3

Table 3. Median mass (kg) for age and height of adolescent girls [WHO, 1985]

Height cm	Age, years								
	10	11	12	13	14	15	16	17	18
120	22.3								
125	24.6	24.7							
130	27.1	27.9	27.3						
135	30.1	30.1	30.7	31.5					
140	32.9	33.1	33.2	34.1	34.8				
145	36.6	36.4	36.6	37.2	39.3	41.4			
150	38.8	40.2	39.9	41.1	43.0	44.6	45.9	46.4	
155		44.0	44.8	45.0	47.0	48.1	50.2	50.4	51.4
160			48.9	49.2	49.8	51.5	51.9	52.8	53.1
165			52.4	53.1	54.0	54.2	54.8	55.4	55.9
170				56.8	57.6	58.0	58.9	58.9	60.1
175					60.0	60.8	61.2	62.1	62.9
180					61.3	62.2	63.0	63.9	64.4

mostly on measurements in industrially developed countries. As mentioned, the latter can be misleading when evaluating the growth of children in developing countries. Tables 1–3, which take into account the proportionality of the organism (also essential to the evaluation of nutritional status), are useful when comparing children living under various conditions in different parts of the world. Extreme malnutrition and/or obesity (that is, a body mass more than ± 2 SD) is nevertheless apparent (table 1).

The body mass index has been accepted as an important measure of morphological development, and is recommended for clinical evaluations, assessment of nutritional status and so forth (body mass, kg/square of the height, m). Most experience of this index has been gained with adults, but growth grids also exist for children [Rolland-Cachera MF, Sempe M: Paris, INSERM, 1985].

Many other body dimensions have been used to evaluate growth: circumferences, lengths and breadth measurements give more detailed information on the development of various parts of the organism and its proportionality. The number and combination of such measurements varies according to the aim of growth studies, the material possibilities, the number of personnel and so forth.

Indices relating various body dimensions (for example, chest circumference to body height, biacromial breadth to body height and/or biiliocristal breadth; waist circumference to maximal hip circumference, that is, the waist/hip ratio) are numerous. The choice among such measures again varies with the aims of a study (for example, the evaluation of sexual maturation, the level of physical fitness and so on).

One of the most widespread studies of human growth and development was performed within the framework of the International Biological Programme (IBP). The outcome of these studies from different parts of the world has been published in recent years.

Body composition has been used as one of the criteria of growth. It also serves as an indicator of both nutritional status and physical fitness. It may be evaluated from various points of view – anatomical, biochemical and functional. Various body compartments have been measured to date, using a variety of methodologies. From the functional point of view, it is now quite common to measure both the absolute and the relative amounts of the two main body constituents – active, lean body mass and depot fat; each of these variables can differ widely, even when considering subjects of similar age, sex, body height and total body mass.

Observations on several thousands of people ranging from newborn infants to the very old have also shown characteristic age and sex differences.

The variability of lean body mass is much smaller than the variation of total body mass. Lean body mass thus provides a better reference standard, and is correlated more significantly with basal metabolic rate, blood volume, aerobic power, vital capacity and so forth.

Many methods have been used to measure body composition. Hydrostatic weighing, with simultaneous measurement of the volumes of air in the lungs and respiratory passages (using the Archimedean principle) has the longest tradition. A further possibility is to calculate the body volume from selected body dimensions (the geometric approach). Finally, predictions can be based on body mass (a statistical approach) which is less reliable. Prediction equations based on circumferences and/or skinfold measurements are also available.

Lean tissue contains a relatively constant proportion of water, so that it can be estimated using various techniques to determine body water (antipyrin, D_2O, T_2O, etc.). The body content of potassium may be determined by several techniques, including the isotopic dilution of injected ^{42}K, or noninvasively by the determination of ^{40}K (whole body counting).

Another approach now being used quite widely is based on differences of electrical conductivity between lean and adipose tissues (bioelectrical impedance); conductivity is substantially greater in the former, due to its higher electrolyte content. This safe and simple noninvasive technique gives results which correlate well with the other methods mentioned above. The same applies to TOBEC method (total body conductivity measurements). During recent years, neutron activation techniques have also become available. These allow an in vivo analysis of body elements; K, Ca and N are of special interest. Computer-assisted tomography (CAT) now allows the estimation of muscle, bone, fat and other tissue proportions at different levels (cuts) within a subject, contributing a new dimension to the evaluation of body composition.

The above-mentioned methods are mainly suitable for laboratory studies. Under field conditions, reliance has been placed upon caliper skinfold measurements, with the estimation of body composition by means of regression equations or tables. The measurement of up to ten skinfolds diminishes the error of predictions, and skinfold data can serve especially for group evaluations. Another approach is to employ regression equations that also include other body dimensions; the results again correlate quite well with data gained by more sophisticated methods.

Physical Fitness and Performance

Fitness is an essential component of health, and may be regarded as a complex of prerequisites allowing an organism to react in an optimal way to various environmental stimuli. Physical movements are a basic manifestation of life, and contribute to the level of physical fitness observed in any given individual. Physical activity has a far-reaching influence on life. Impaired physical activity often appears as a result of either a reduced or a too abundant food supply. It has an impact on the functional development of the human organism, and is an important factor contributing to deviations from normal growth patterns. Further, lack of physical activity plays a role in the pathogenesis of some diseases that only become manifest in later life.

Physical fitness can be assessed by measuring the functional capacity of the various body systems involved. Fitness implies optimal reactions to demanding situations – both heavy work-loads and accompanying environmental stresses. However, the assessment of physical fitness is frequently restricted to an evaluation of the efficiency of circulatory and respiratory systems, to an evaluation of motor abilities, or to a determination of total work output, which under special situations may depend primarily upon either muscular strength or cardiorespiratory endurance. Thus, the interpretation of nutritional and other environmental stimuli varies considerably, depending on the approach used to evaluate physical fitness. Plainly, a complex and well-integrated evaluation of physical fitness awaits substantial methodological development.

When correlations are sought between diet and fitness, the individual aspects of physical fitness must be differentiated. Along with the morphological prerequisites mentioned above, physical fitness and performance are determined mainly by peak energy output (a summation of aerobic and anaerobic processes, depending largely on the efficiency of the cardiorespiratory system), with other contributions from neuromuscular function (strength and technique) and from psychological factors (particulary motivation). Individual factors play a more or less dominant role, depending on the nature of the physical task.

The various factors are modified in different ways by diet and exercise, whether excessive or inadequate. Heavy physical work, for example exercise demanding a high percentage of endurance activity, contributes greatly to the development of cardiorespiratory efficiency. Strength exercises and static work, on the other hand, contribute to the development of muscle strength, while skill training develops coordination. All of these changes improve overall physical fitness, but gains in one area can occur without stimulating the other aspects of fitness mentioned above.

The interplay of the above-mentioned factors must be considered relative to quite widely differing implications of physical fitness: first of all, there are health aspects such as the achievement of 'positive health' and the prevention of cardiovascular and/or metabolic diseases, particularly in the industrially developed countries. There are issues of improving economic productivity, especially in non-industrialized societies dependent on physical strength, endurance and physical skill. Last but not least, there is a concern to achieve the highest possible performance in competitive sports.

From all these points of view, it is essential to consider the individual items of physical fitness both separately and synthetically, having in mind the added impact of environmental factors, including nutrition.

When analysing the impact of nutrition relative to exercise and muscular work, it is important to take into account the age of the individual and the stage of development. It must be decided whether the factors under review began to influence the organism in early childhood, in adolescence or as an adult. The consequences may be either temporary or permanent, depending on the intensity and duration of the stimulus. Brief stimuli are usually of lesser importance for both the current and the future development of the organism, particularly if they are experienced during the later periods of development. Long-lasting stimuli modify development in a much more important way, whether the change be nutritional or an altered physical demand.

The character and intensity of the stimuli are other factors which influence adaptive changes of morphological and functional status.

The ability to perform physical work depends first of all upon the capacity of the cardiovascular and respiratory systems, as characterized by aerobic power. The ability to carry oxygen to the working tissues – particularly the active muscles, depends on the efficiency of the cardiorespiratory system. It can be evaluated by measuring the oxygen uptake at a given power output, along with necessary factors such as heart and ventilation rate, the respiratory minute volume, the carbon dioxide output or the respiratory gas exchange ratio (R). The maximal oxygen intake may be expressed in either absolute or relative terms (for instance, per kg of body mass or lean body mass, or per heart beat). A large aerobic power is characteristic of a physically fit individual; this is especially true of the ability to perform dynamic activities such as running, which parallel a good overall level of health, and an adequate or good nutritional status.

The maximal oxygen intake is measured on a treadmill or a cycle ergometer. Some authors prefer the latter device because it allows a more precise calculation of power output. Nevertheless, the results obtained from a cycle ergometer can be influenced by prior experience in cycling. Particularly in underdeveloped societies, running is a more natural physical activity. Despite efforts at international standardization of technique, very different methodological approaches remain. In some laboratories, subjects are tested on a treadmill (with or without a variation of slope) during a graded work test to subjective exhaustion. The initial choice of treadmill speed depends on the fitness of the subject to be tested, being lower for inactive individuals than for athletes. When procedures differ between various laboratories, it becomes difficult to compare the results obtained on individual populations. Fortunately, this problem was avoided in studies under the aegis of the International Biological Programme (IBP), where internationally agreed methods were used, including PWC_{170} (physical working capacity at the pulse frequency 170/min).

From the respiratory parameters measured during a maximal effort test, the ventilatory threshold can be determined, either using blood lactate as a criterion, or non-invasively from the dependence of pulmonary ventilation on the oxygen uptake and/or carbon dioxide output, using two-part discontinuous linear models.

Under field conditions, the step test has the important advantage of simplicity, but there may be difficulties in taking the necessary physiological measurements. Single, double, triple and multiple steps have been proposed. The single and double steps have found particular application in large population groups. The frequency of stepping depends on age and fitness level, as well as the height of the step. Standardized procedures have been suggested: heart rates are counted at rest, while climbing steps and during recovery, and various indices are calculated such as the step test index and the cardiac efficiency index.

Muscle strength is assessed from the maximal force in Newtons exerted by a given muscle group (for example, the flexors and extensors of the elbow and the knee, the extensors of the trunk, the plantar flexors, and the hand grip force), contractions being developed under isometric conditions in a standard position. Either an electric or a mechanical dynamometer can be used.

Motor performance can be evaluated from the results of field tests such as the 50- and 100-metre dash, broad jump and soft ball throw. The test battery of AAHPERD (American Association of Health, Physical Education, Recreation and Dance) is widely used; this includes pull-ups, sit-ups, shuttle run, standing broad jump, and 600 yard walk-run. Other possibilities are described in specialized textbooks [e.g. Eurofit. Rome, 1988].

As regards malnutrition, the aerobic power can deteriorate due to an inadequate food intake, but it is also low relative to total and lean body mass in obese subjects. The same applies to step test results. Nevertheless, if malnutrition is marginal, the results of the above-mentioned tests may be quite satisfactory and comparable to the results observed in well-nourished subjects.

Muscle strength correlates closely with lean body mass and the results of other physical performance tests. Changes of both absolute and relative muscle strength may result from malnutrition. Very obese subjects sometimes have an advantage of strength, as shown by extreme cases such as Japanese sumo wrestlers and champion weightlifters.

Nutritional Status and Dietary Intake

As with criteria for growth, somatic development, physical fitness and performance, it is also difficult to define the optimal nutritional status and dietary intake for the all-round development of humans. Dietary allowances have been recommended in many countries [WHO, 1985], but as observed in experimental studies, individuals with a similar morphological development and way of life (including athletes and/or individuals from the developing countries with marginal malnutrition) may show twofold differences of energy intake. Obviously, energy efficiency must vary widely, enabling humans to function under very different conditions of energy input and output. 'how much is enough'... or 'how much is too little or too much' has remained a problem, especially when considering the wide range of above-basal energy output due to spontaneous and induced physical activity.

From the historical point of view, the intake of energy and of individual food components (particularly proteins) has been changing over the centuries. Indirect evidence is available from historical records in Europe. During periods of relative affluence, the intake of food has been rather high – 750 g of bread and 250 g of meat per day. On the other hand, during less favourable periods it has dropped to 250 g of bread and 40 g of meat per day. Health and performance problems have obviously been related to these changes of food intake. Nevertheless, most humans have survived and remained productive during all of these periods.

When considering the RDA, it is necessary to keep in mind the 'requirement for what'. There remain many unknown factors producing interindividual variations. Sex, age and professional workloads inevitably influence energy needs, as follows from the spontaneous energy intake and selection of individual nutrients. The usual descriptor of energy needs is the average of individual requirements in a particular class of subjects, as in the case of other standards such as growth and fitness.

In the case of protein requirements, a safe level of intake is considered, one that will meet or exceed the requirements of practically all individuals, explicitly taking into account individual variations in requirements; this safe level is defined as the average requirement $+2\,SD$. In addition to genetic interindividual differences, adaptation can change the need for both energy and individual nutrients. Adaptation may here be defined as a dynamic process by which a new steady state is achieved in response to environmental stimuli.

Table 4. Equations for predicting basal metabolic rate (BMR) from body mass (M) in growing males and females [WHO, 1985]

Age range years	kcal/day	SD	MJ/day	SD
Males				
0–3	60.9 M − 54	53	0.255 M − 0.228	0.222
3–10	2.7 M + 495	62	0.0949 M + 2.07	0.259
10–18	17.5 M + 651	100	0.0732 M + 2.72	0.418
Females				
0–3	61.0 M − 51	61	0.255 M − 0.214	0.255
3–10	22.5 M + 499	63	0.0941 M + 2.09	0.264
10–18	12.2 M + 746	117	0.0510 M + 3.12	0.489

A true steady state can never be achieved – life implies a dynamic balance with the environment with respect to energy input and output as well as their balance and turnover. Adaptation has various aspects – metabolic, functional, morphological, social and behavioural, which become manifest under the varying conditions of life encountered around the world.

It has often been suggested that a reduction of body size might be one such manifestation – a 'useful' adaptation to limited supplies of energy and protein; however, this is only true if other bodily functions are not influenced adversely.

The most recent consensus on energy and protein needs reached at the Joint FAO/WHO/UNU Expert Consultation was later confirmed on the occasions of the 13th (1985, Brighton, UK) and 14th International Congresses of Nutrition (1989, Seoul, South Korea). Recommended allowances were based on the basal metabolic rate (BMR), as estimated from the total body mass and height, and/or only from body mass (table 4). Needs beyond the basal energy requirements were established as multiples of basal metabolic rate, according to the daily work-loads [WHO, 1985].

Calculations of basal metabolic rate (BMR) and total energy expenditure for boys and girls aged 10.5 to 17.7 years of age are presented in tables 5 and 6. More specific examples of the calculation of daily energy requirements for a 10.5-year old boy and girl are given in tables 7 and 8. The safe levels of protein intake for adolescent girls and boys are given in table 9.

Recent experience has led to a downward adjustment of the recommended energy intake; comparing values for 1971 and 1981, there has been a reduction of 16–20% in males, and 18–20% in females aged 10–18

Table 5. Calculation of BMR and total energy expenditure [WHO, 1985]

Boys	Age, years							
	10.5	11.5	12.5	13.5	14.5	15.5	16.5	17.5
Mass, kg	32.2	37.0	40.9	47.0	52.6	58.0	62.7	65.0
BMR/day[1]	**1,215**	**1,299**	**1,367**	**1,474**	**1,572**	**1,666**	**1,748**	**1,789**
	5,084	5,435	5,720	6,167	6,577	6,971	7,314	7,485
BMR/min	**0.844**	**0.900**	**0.949**	**1.034**	**1.092**	**1.157**	**1.214**	**1.242**
	3.530	3.764	3.972	4.283	4.568	4.841	5.079	5.198
Calculation of energy expenditure								
BMR	*24h*	*24h*	*24h*	*24h*	*24h*	*24h*	*24h*	*24h*
	1,215	**1,299**	**1,367**	**1,474**	**1,572**	**1,666**	**1,748**	**1,789**
	5,084	5,435	5,720	6,167	6,577	6,971	7,314	7,485
School(+0.6 BMR)	*4h*	*5h*	*5h*	*5h*	*6h*	*6h*	*6h*	*6h*
	121	**162**	**171**	**184**	**236**	**250**	**262**	**268**
	508	679	715	771	987	1046	1097	1123
Light activity (+0.6 BMR)	*4h*	*5h*	*6h*	*7h*	*7h*	*7h*	*7h*	*7h*
	121	**130**	**171**	**221**	**275**	**292**	**306**	**313**
	508	544	715	925	1,151	1,220	1,280	1,310
Moderate activity (+1.5 BMR)	*6.5h*	*5.5h*	*4.5h*	*3.5h*	*2.5h*	*2.5h*	*2.5h*	*2.5h*
	494	**466**	**384**	**332**	**246**	**260**	**273**	**280**
	2,065	1,868	1,609	1,349	1,028	1,089	1,143	1,170
High activity (+6.0 BMR)	*0.5h*	*0.5h*	*0.5h*	*0.5h*	*0.5h*	*0.5h*	*0.5h*	*0.5h*
	127	**135**	**142**	**153**	**163**	**173**	**182**	**186**
	530	566	596	642	684	726	762	780
Daily growth[2]	**62**	**71**	**78**	**90**	**101**	**55**	**30**	**31**
	258	296	327	376	421	232	125	130
Total daily energy expenditure	**2,140**	**2,244**	**2,314**	**2,445**	**2,592**	**2,697**	**2,801**	**2,867**
	8,953	9,388	9,681	10,230	10,847	11,283	11,721	11,997
Daily energy expenditure/ kg body mass	**66.5**	**60.6**	**56.6**	**52.0**	**49.3**	**47.0**	**44.7**	**44.1**
	278.0	253.7	236.7	217.7	206.1	195.0	186.9	184.6

The figures in *italics* give the daily duration of the activity in question: the balance, of up to 24 h is made up by the period of sleep (1.0 BMR). The figures in **bold type** give the amount of energy in $kcal_{in}$ and those in ordinary type (unless otherwise stated) the amount of energy in kJ.

In calculating the total daily energy expenditure the BMR is applied to the whole 24 h, additional amounts being added for the various specified types of activity, e.g. for boys aged 10.5 years. +0.6 BMR for 4 h of light activity.

[1] $BMR(kcal_{in}/day)$: for boys = 17.5 M + 651: for girls = 12.2 M + 746.
[2] The energy expenditure on growth was taken as 8kJ/kg of body mass at 10–15 years: 4 kJ/kg at 15 years: and 2 kJ/kg at 16–18 years.

Table 6. Calculation of BMR and total energy expenditure [WHO, 1985]

Girls	Age, years							
	10.5	11.5	12.5	13.5	14.5	15.5	16.5	17.5
Mass, kg	33.7	38.7	44.0	48.8	51.4	53.0	54.0	54.4
BMR/day[1]	**1,157**	**1,218**	**1,282**	**1,341**	**1,373**	**1,393**	**1,405**	**1,410**
	4,841	5,096	5,364	5,611	5,745	5,828	5,879	5,899
BMR/min	**0.804**	**0.846**	**0.890**	**0.931**	**0.953**	**0.967**	**0.976**	**0.979**
	3.362	3.539	3.725	3.897	3.989	4.047	4.082	4.097
Calculation of energy expenditure								
BMR	*24h*	*24h*	*24h*	*24h*	*24h*	*24h*	*24h*	*24h*
	1,157	**1,218**	**1,282**	**1,341**	**1,373**	**1,393**	**1,405**	**1,410**
	4,841	5,096	5,364	5,611	5,745	5,828	5,879	5,899
School (+0.5 BMR)	*4h*	*5h*	*5h*	*5h*	*6h*	*6h*	*6h*	*6h*
	96	**127**	**133**	**140**	**172**	**174**	**175**	**176**
	403	531	558	585	718	729	734	738
Light activity (+0.5 BMR)	*4h*	*4h*	*5h*	*6h*	*7h*	*7h*	*7h*	*7h*
	96	**102**	**133**	**168**	**200**	**203**	**205**	**206**
	403	425	558	702	838	851	857	861
Moderate activity (+1.2 BMR)	*6.5h*	*5.5h*	*4.5h*	*3.5h*	*2.5h*	*2.5h*	*2.5h*	*2.5h*
	376	**335**	**288**	**235**	**172**	**174**	**175**	**176**
	1572	1402	1205	983	719	729	734	738
High activity (+5.0 BMR)	*0.5h*	*0.5h*	*0.5h*	*0.5h*	*0.5h*	*0.5h*	*0.5h*	*0.5h*
	120	**127**	**140**	**140**	**143**	**145**	**146**	**147**
	504	531	588	585	599	608	612	615
Daily growth[2]	**64**	**74**	**84**	**93**	**98**	**51**	**26**	**26**
	270	310	352	390	411	212	108	109
Total daily energy expenditure	**1,910**	**1,982**	**2,054**	**2,117**	**2,158**	**2,140**	**2,133**	**2,142**
	7,993	8,294	8,595	8,856	9,029	8,956	8,924	8,960
Daily energy expenditure/ kg body mass	**56.7**	**51.2**	**46.7**	**43.4**	**42.0**	**40.4**	**39.5**	**39.4**
	237.2	214.3	195.3	181.5	175.7	169.0	165.3	164.7

The figures in *italics* give the daily duration of the activity in question; the balance, up to 24 h is made up by the period of sleep (1.0 BMR). The figures in **bold type** give the amount of energy in $kcal_{in}$ and those in ordinary type (unless otherwise stated) the amount of energy in kJ.
In calculating the total daily energy expenditure, the BMR is applied to the whole 24 h, additional amounts being added for the various specified types of activity, e.g. for boys aged 10.5 years, +0.6 BMR for 4 h of light activity.
[1] BMR ($kcal_{in}$/day): for boys = 17.5 M + 651; for girls = 12.2 M + 746.
[2] The energy expenditure on growth was taken as 8 kJ/kg of body mass at 10–15 years; 4 kJ/kg at 15 years; and 2 kJ/kg at 16–18 years.

Table 7. Example of the calculation of the daily energy expenditure of a 10.5-year-boy in a developing country (body mass = 33.2 kg) [WHO 1985]

Activity	h	kcal$_{th}$	kJ
Sleep at 1.0 × BMR[1]	9	455	1,900
School at 1.6 × BMR	2.5	200	840
Light activity at 1.6 × BMR			
Sitting, standing, moving around	6.5	525	2,200
Social activities, washing, play	2	160	670
Moderate activity at 2.5 × BMR			
Walking, household tasks, agricultural tasks, play	3	380	1,590
Heavy activity at 6.0 × BMR			
Fetching wood and water, agricultural tasks	1	300	1,260
Growth		60	250
Total requirement per 24 h = 1.71 × BMR		2,080	8,710

[1] BMR estimated to be 1,215 kcal$_{th}$/day (5,080 kJ/day).

Table 8. Example of the calculation used to derive energy expenditure in a 10.5-year-old girl (body mass = 33.8 kg) [WHO, 1985]

	h	kcal$_{th}$	kJ
Sleep at 1.0 × BMR[1]	9	435	1,820
School at 1.5 × BMR	4	290	1,210
Light activity at 1.5 × BMR	4	290	1,210
Moderate activity at 2.2 × BMR	6.5	690	2,890
High activity at 6.0 × BMR	0.5	145	610
Total expenditure		1,850	7,740
Growth		65	270
Total requirement per 24 h = 1.65 × BMR		1,915	8,010

[1] BMR estimated to be 1,160 kcal$_{th}$/day (4,850 kJ/day).

years. The reduction is relatively larger for older than for younger males, but the situation is reversed in the females.

Studies of English infants show that over the first year of life, growth and development has remained normal despite a reduction of the local RDA [WHO, 1985]. The new procedures for the estimation of energy requirements using a combination of BMR and energy output enable nutritionists to individualize energy needs, avoiding unnecessarily high

Table 9. Safe level of protein intake for adolescent girls and boys (10–18 years) [WHO, 1985]

Age Years	Maint-enance mg N/kg/day	Growth mg N/kg/day	Total mg N/kg/day	+2 SD mg N/kg/day	Safe level g protein/kg/day	1971 Committee g protein/kg/day
Girls						
10–11	110	19	129	161	1.00	0.78
11–12	109	17	126	157	0.98	0.75
12–13	108	15	123	154	0.96	0.71
13–14	107	13	120	150	0.94	0.65
14–15	106	9	115	144	0.90	0.60
15–16	105	7	112	140	0.87	0.58
16–17	104	2	106	132	0.83	0.57
17–18	103	0	103	129	0.80	–
Boys						
10–11	110	17	127	159	0.99	0.82
11–12	109	17	126	157	0.98	0.80
12–13	108	21	129	161	1.00	0.78
13–14	107	17	124	155	0.97	0.75
14–15	106	17	123	154	0.96	0.70
15–16	105	13	118	147	0.92	0.65
16–17	104	11	115	144	0.90	0.63
17–18	103	7	110	137	0.86	–

energy intakes that not only do not help the developmental process, but may also initiate pathological changes and impose heavy economic burdens on poor societies.

Six main age ranges are specified for the most recent table of recommended dietary allowances, allowing a differentiation according to the developmental stage. From 10 years of age, there are separate RDAs for the two sexes. Many developed and developing countries have their own local RDA, although this often corresponds with the values cited above. Nevertheless, recommendations have as yet allowed little individualization for differing daily physical work-loads.

Catch-up growth after a period of malnutrition is a special problem. The subject may fall below an acceptable range of body mass for height ('wasting') or body size may be small because the increase in height does not meet accepted standards ('stunting'). The former case needs of energy and protein for catch-up growth have been calculated in some studies on the rehabilitation of malnourished children (for example, 5 kcal or 21 KJ and 0.23 g protein per g of tissue laid down). The daily increments needed depend on the rate at which catch-up growth is to be achieved, but it is clear that at all rates of catch-up, there must be some increase in the ratio of protein to energy requirement, over and above the RDA.

An interesting problem would be the catch-up of functional capacity, of physical performance and fitness during refeeding after a period of malnutrition. In some studies it has been shown that an appropriate physical activity regime, along with the refeeding can speed up both convalescence and catch-up growth of the children, especially from the point of view of lean body mass development. In spite of the fact that under such conditions children generally remain smaller, it seems to be of much greater importance to have them fit and healthy, than to achieve a large body size.

Jana Pařízková, MD, Research Institute for Physical Education,
Charles University, (VOT UK), Ujezd 450, 11807 Prague 1 (Czechoslovakia)

Growth, Exercise, Nutrition and Fitness in China

Ji Di Chen

Institute of Sports Medicine, Beijing, People's Republic of China

The increase of height and body mass among Chinese students throughout the period of growth and development is summarized in table 1. Data are also available for aerobic power over the adolescent years; these show relatively constant average values of 50–53 ml/kg · min for male students (table 2).

Many factors contribute to the growth and development of functional capacity, physical exercise and good nutrition being particularly important to sound growth and health. A regular, optimal exercise programme enhances metabolism, accelerates growth and development, enhances function, and increases the physical working capacity. In contrast, limited activity is associated with excessive body mass and fat accumulation, while physical fitness, resistance to disease and health prognosis do not reach desirable levels [13]. A well-planned diet not only contributes to current growth and development, but also has positive effects upon adult physique and health. The impact of malnutrition is much more serious during the period of growth than in adult life. In addition to interfering with growth and development, there is an impairment of immune function that leads to an increased incidence of disease and a higher mortality rate. Nevertheless, there is still need for a more precise definition of optimal patterns of nutrition and exercise for the growing child.

Effects of Exercise on Growth, Body Function and Physical Work Capacity

Rate of Growth and Development

Systematic and comprehensive exercise training accelerates the growth of children and adolescents. For example, if the average height and body mass of adolescent soccer players is compared with other students in the

Table 1. Average height and body mass of Chinese adolescents at various ages [data taken from ref. 1]

Age, years	8	9	10	11	12
Body mass, kg					
Boys	23.2 ± 2.77	25.5 ± 3.18	28.0 ± 3.67	30.5 ± 4.06	34.0 ± 5.06
Girls	22.5 ± 2.74	24.9 ± 3.24	27.8 ± 4.03	31.0 ± 4.81	35.4 ± 5.44
Height, cm					
Boys	125.7 ± 5.32	130.6 ± 5.59	135.3 ± 5.75	139.9 ± 5.98	145.2 ± 6.80
Girls	125.0 ± 5.36	130.1 ± 5.72	135.6 ± 6.31	141.2 ± 6.71	147.1 ± 6.60

Table 2. Maximal oxygen intake and peak heart rate of male Chinese students ($\bar{x}$ ± SD; data taken from Zhou and Lu [23])

Age years	n	Height cm	Body mass kg	$\dot{V}_{O_2 max}$ l/min	$\dot{V}_{O_2 max}$ ml/kg·min	HR_{max} beats/min
13.5 ± 0.3	32	156.4 ± 8.2	40.9 ± 7.5	2.06 ± 0.42	50.6 ± 6.6	193.3 ± 9.0
14.5 ± 0.3	31	165.3 ± 6.6	50.5 ± 8.3	2.51 ± 0.28	50.2 ± 4.7	191.9 ± 7.2
15.3 ± 0.3	32	167.3 ± 5.9	53.5 ± 7.9	2.75 ± 0.37	51.9 ± 6.4	191.9 ± 9.0
16.4 ± 0.3	32	170.7 ± 5.6	54.2 ± 6.6	2.75 ± 0.35	50.8 ± 4.9	188.0 ± 7.9
17.5 ± 0.3	34	169.2 ± 6.2	56.4 ± 4.6	2.93 ± 0.38	52.0 ± 5.3	187.6 ± 7.6
18.4 ± 0.3	32	168.7 ± 6.8	57.4 ± 6.0	3.00 ± 0.40	52.5 ± 4.6	185.2 ± 8.8
19.3 ± 0.3	33	172.7 ± 5.8	60.9 ± 6.7	3.20 ± 0.40	52.3 ± 4.0	185.5 ± 7.8

Table 3. Growth of height and body mass in amateur soccer players and control students [data from refs. 1, 18]

Age years	Height, cm				Body mass, kg			
	mean value		annual increase		mean value		annual increase	
	soccer players	students	soccer players	students	soccer players	students	soccer players	students
10	139.1	134.4	–	–	30.14	27.4	–	–
11	143.7	140.4	4.66	6.0	32.53	30.6	2.40	3.20
12	149.7	146.0	5.94	5.6	36.35	34.3	3.81	3.70
13	157.0	152.4	7.99	6.4	42.24	38.6	5.89	4.30
14	164.2	158.7	7.15	6.3	48.68	43.8	6.44	5.20
15	168.2	164.3	4.00	5.6	53.76	48.8	5.08	5.00
16	171.6	167.9	3.41	3.6	56.99	52.4	3.23	3.60
17	173.1	170.1	1.52	2.1	59.93	54.5	2.94	2.10

13	14	15	16	17	18
38.6 ± 6.44	44.1 ± 6.86	49.0 ± 6.64	52.5 ± 6.14	54.8 ± 5.75	56.5 ± 5.49
39.8 ± 5.88	43.5 ± 5.75	46.4 ± 5.57	48.3 ± 5.51	49.2 ± 5.49	50.7 ± 5.50
151.8 ± 8.03	158.3 ± 7.84	163.8 ± 6.87	167.0 ± 5.94	168.6 ± 5.72	169.3 ± 5.60
151.6 ± 6.16	154.8 ± 5.59	156.8 ± 5.33	157.8 ± 5.22	158.1 ± 5.18	158.2 ± 5.08

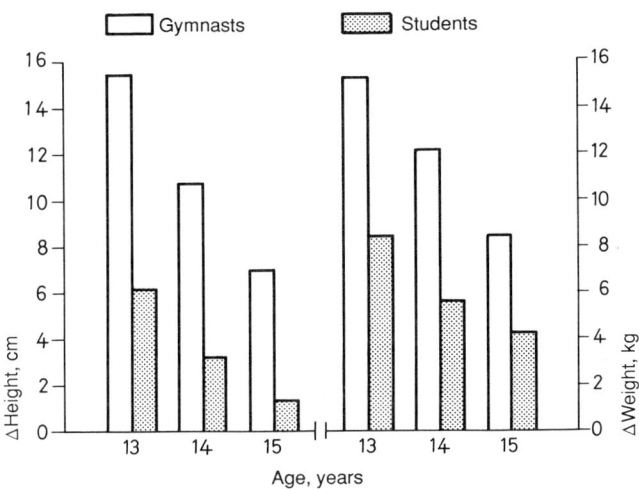

Fig 1. Comparison of the annual increase of height and body mass between gymnasts and control students [from ref. 3].

same city, it is plain that the height and body mass at a given age are greater in athletes who are undergoing systematic training for amateur competition (table 3) [18]. A 3-year follow-up of 11 elite girl gymnasts shows a similar pattern, the gymnasts growing faster than other female students in the same city. However, in this second instance, the interpretation is complicated because the body size of the gymnasts was much smaller than the average students at the beginning of the study; the growth of the gymnasts might thus be delayed, particularly if their small initial size was not genetically determined. However, it is probable that there is some contribution again from exercise training and well-planned nutrition (fig. 1) [3].

Table 4. A comparison of cardiorespiratory function between adolescent runners and control students [material taken from ref. 10]

Subject	Sex	n	Exercise load	$\dot{V}_{O_2}$ litres/min	$\dot{V}_{O_2\,max}$ litres/min	$\dot{V}_{O_2\,max}$ ml/kg·min	HR beat/min	O_2·Pulse ml/beat
Runners	M	26	preload	0.647 ± 0.172**	–	–	60.5 ± 11.8**	11.05 ± 3.11**
			during load	3.608 ± 0.410**	3.750 ± 0.669**	57.2 ± 9.3**	165.0 ± 13.9*	22.11 ± 3.54**
Nonathlete controls	M	52	preload	0.525 ± 0.134	–	–	73.7 ± 10.9	7.33 ± 2.21
			during load	2.440 ± 0.604	2.592 ± 0.583	46.2 ± 12.9	173.0 ± 12.9*	14.10 ± 4.23
Runners	F	17	preload	0.505 ± 0.131**	–	–	71.9 ± 12.1*	7.32 ± 2.33*
			during load	2.616 ± 0.243**	2.797 ± 0.329**	49.3 ± 4.8**	185.1 ± 6.3	14.30 ± 1.49**
Nonathlete controls	F	56	preload	0.409 ± 0.101	–	–	79.8 ± 11.8	5.24 ± 1.56
			during load	1.950 ± 0.401	1.937 ± 0.378	39.8 ± 8.7	177.2 ± 11.8	10.88 ± 2.10

Data compared with that of the nonathlete controls showed significance: ** $p < 0.01$, * $p < 0.05$.
Exercise load: exercise was carried out on a cycle ergometer in sitting position for 6 min.
Exercise load for runners: male: 1,800 kpm/min; female: 1,500 kpm/min.
Exercise load for nonathlete controls: male: 1,500 kpm/min, female: 1,200 kpm/min.

Cardiopulmonary Function

A comparison between adolescent runners (average age 17.3 ± 2.8 years) and nonathletic controls shows significantly higher values for both maximal oxygen intake and oxygen pulse in the athletic students (table 4). Ultrasound cardiographic comparisons between elite adolescent soccer players and other students matched for age demonstrate that physical exercise training is associated with both structural and functional changes in the heart. The ventricular dimensions are significantly larger in the athletes, these changes being particularly marked in those with a history of prolonged and systematic training (tables 5, 6) [16].

Table 5. A comparison of cardiac morphology between soccer players and nonathletic controls [data taken from ref. 16]

Subject groups	Age years	n	LVDd, mm	IVST, mm
Nonathlete controls	14	37	45.20 ± 3.70	6.80 ± 0.70
Soccer players	14	24	49.52 ± 2.13*	7.25 ± 0.54
Nonathlete controls	15	30	45.60 ± 3.40	7.20 ± 0.80
Soccer players	15	80	47.59 ± 2.81	7.33 ± 0.52
Nonathlete controls	16	32	46.50 ± 3.50*	7.80 ± 0.80
Soccer players	16	97	48.65 ± 2.88	7.46 ± 0.56

LVDd = Left ventricular diastolic diameter; IVST = interventricular septal thickness.
Data of the controls compared with that of the athletes showed significance: *$p < 0.01$.

Table 6. The influence of training duration upon the development of ventricular hypertrophy [data taken from ref. 16]

Items	Training time	
	<4 years (n = 116)	>4 years (n = 85)
LVDd, mm	47.43 ± 3.14**	48.68 ± 2.44**
LVDs, mm	30.50 ± 2.38*	31.27 ± 2.24*
IVST	6.02 ± 0.47**	6.22 ± 0.55**
SV	79.19 ± 16.16**	84.87 ± 12.65**
LVV	107.72 ± 21.45**	116.10 ± 17.15**

LVDd = Left ventricular diastolic diameter; IVST = interventricular septal thickness; LVDs = left ventricular systolic diameter; SV = stroke volume; LVV = left ventricular volume.
*$p < 0.05$; **$p < 0.01$.

Table 7. A comparison of physical characteristics between physical education students and those enrolled in other programmes [data taken from ref. 22]

	College students	Physical education students
n	50	40
Height, cm	173.8 ± 4.2*	176.4 ± 4.5
Body mass, kg	61.7 ± 4.5**	66.4 ± 5.2
Body fat, %	11.76 ± 3.8*	9.83 ± 2.7
Lean body mass, kg	54.3 ± 3.78**	59.7 ± 3.99

Data compared with that of the physical education students showed significance: *p < 0.01; **p < 0.001.

Table 8. The basal metabolic rate (kcal/m^2/h) in relation to age [data taken from ref. 19]

	Age, years										
	1	3	5	7	9	11	13	15	17	19	20
Male	53.0	51.3	49.3	47.3	45.2	43.0	42.3	41.8	40.8	39.2	38.6
Female	53.0	51.2	48.4	45.4	42.8	42.0	40.3	37.9	36.3	35.5	35.3

Body surface (m^2) = 0.0061 height (cm) + 0.0128 weight (kg) − 0.1529.

Body Composition

Exercise participation is associated with a smaller percentage of body fat and higher lean body mass values, both of these differences contributing to a greater physical work capacity [14, 15]. Again, the differences are most marked in athletes reaching high levels of training (table 7, fig. 2, 3) [3, 6, 8, 22].

Effects of Exercise Programmes upon the Nutritional Requirements of Children and Adolescents

Energy Requirements

The energy requirement of children and adolescents has four components – basal metabolism, specific dynamic action, exercise activities, and growth.

Basal Metabolism. The basal metabolic rate (BMR) per unit of body surface area is higher for children and adolescents than for adults (table 8) [19]. Expressing data in kcal (1 kcal = 4.186 kJ), the adult BMR is about

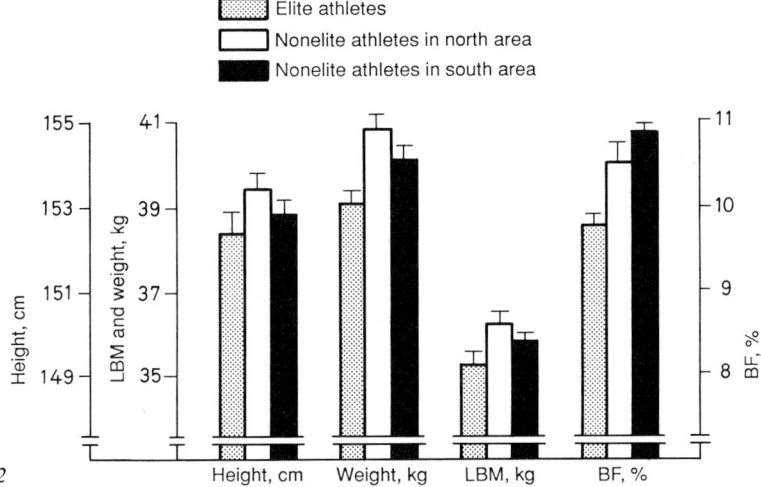

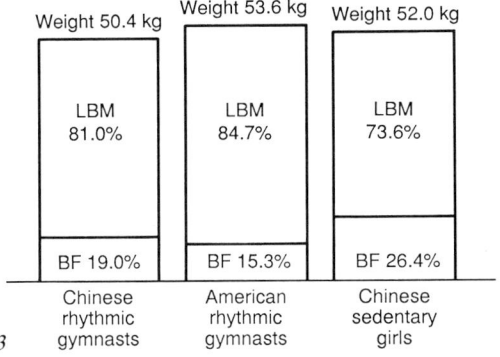

Fig. 2. Comparison of body composition between elite and nonelite athletes [from ref. 3].

Fig. 3. Comparisons of body composition between gymnasts and sedentary young girls. Based on data of Gao et al. [8].

1,440 kcal/day, or 35.2–37.5 kcal/m^2/h, but for children aged 7 years, the corresponding figures are 47.3 kcal/m^2/h in males and 45.4 kcal/m^2/h in females, while for 13-year-old adolescents, the values are 42.3 kcal/m^2/h in males and 40.3 kcal/m^2/h in females. Relating values to body mass, the average is 24 kcal/kg/day in adults, but 44 kcal/kg/day at 7 years, and 30 kcal/kg/day at 13 years.

Specific Dynamic Action (SDA). The SDA normally amounts to some 10% of BMR, but in children and adolescents undergoing systematic physical training, the SDA apparently rises to 15% of BMR, because the energy from protein sources is increased, and gastrointestinal function is inhibited during exercise.

Energy Cost of Exercise. The energy cost of exercise varies with intensity, frequency and duration of training sessions. There are also interindividual differences and adaptations as training proceeds. Typically, amateur physical training demands 100–200 kcal/h, so that 2–4 h of training imposes a demand in the range 200–1,000 kcal [2, 5].

Growth and Development. The energy required for growth and development depends upon the rate of growth. The WHO proposal is to allow 8 kJ/kg for students aged 10–15 years, 4 kJ/kg for those aged 15 years, and 2 kJ/kg for those aged 16–18 years [20].

Either an excess or a deficiency of energy is detrimental to health. If obesity develops during childhood or infancy, this increases the risk of coronary disease in adult life. The WHO recommendations are compared with the Chinese recommendations for children and adolescents in table 9 [9]. Note that the RDA is substantially larger for boys than for girls at all ages.

Table 9. Recommended intake of energy and protein for adolescents aged 10–18 years: a comparison of WHO and Chinese RDAs [data taken from ref. 9]

Age years	Daily energy requirements, kcal/day		Level of protein intake, g/kg	
	WHO	China	WHO	China
Boys				
10–12[1]	2,200	2,300	1.00	1.5–2.0
12–14	2,400	–	1.00	–
14–16[2]	2,650	2,600	0.95	1.7
16–18	2,850	3,000	0.90	1.7
Girls				
10–12[1]	1,950	2,300	1.00	1.5–2.0
12–14	2,100	–	0.95	–
14–16[2]	2,150	2,500	0.90	1.8
16–18	2,150	2,700	0.80	1.7

[1] Data for Chinese boys and girls aged 10–13 years.
[2] Data for Chinese boys and girls aged 13–16 years.

Protein Requirement. The Chinese Nutrition Society has suggested that protein should account for at least 12% of energy intake, corresponding to 70–90 g/day for boys and 75–80 g/day for girls, or 1.5–2.0 g/kg body mass. These figures exceed the WHO recommendation. The protein requirements of children and adolescents undergoing systematic exercise training are further increased, to 3 g/kg at 9–11 years and 2 g/kg at 12–14 years [4]. In setting these high standards, account was taken of the fact that as much as two-thirds of the protein available to Chinese students was from low quality sources. The protein requirement is influenced by both the quality (amino acid composition) and also the overall nutrient status of the subject. The nitrogen balance quickly becomes negative if the overall supply of energy is deficient, whether this situation arises from poor nutrition or intensive exercise. For these reasons, we recommend that adolescent Chinese athletes be given protein to 15% of the total energy expenditure [4].

Mineral Requirements. If temperatures are moderate, the mineral requirements differ little between active and inactive students. The Chinese Nutrition Society recommends larger intakes of Ca and Fe than the WHO RDA (table 10). Many factors are involved in the absorption and utilization of these minerals, so attention must be focussed upon both nutritional balance and food sources of Ca and Fe. Adolescent athletes gain only 1/3 to 1/2 of their Ca from milk, and less than 10% of Fe from heme [2, 5]. Thus, the absorption and utilization would be poor relative to that encountered in some other countries.

Vitamin Requirements. Vitamins are essential not only to growth and development, but also to neural function and the development of acquired immunity. The RDA of vitamins, as suggested by the Chinese Nutrition Society, is again greater than that of the WHO (table 10), although nutritional surveys of Chinese athletes have indicated that these standards are often not met. Vitamin B_{12} and folic acid requirements have not been specified in China. Tentative standards for athletes have been set, based on nutrition surveys and vitamin loading tests (table 11) [17]; the requirements of the athletes are greater than those of sedentary students.

Nutritional Problems of Chinese Amateur Child and Adolescent Athletes

Energy Intakes
Several recent nutritional surveys have shown that most male athletes except gymnasts have an adequate intake of energy; in general, intakes

Table 10. Recommended intake of selected nutrients: comparison of WHO and Chinese RDAs [data taken from ref. 27]

Nutrient	RDA	Children	Adolescents (girls)			Adolescents (boys)		
		7–9 years	10–12 years	13–15 years	16–19 years	10–12 years	13–15 years	16–19 years
Ca, g	WHO	0.4–0.5	0.6–0.7	0.6–0.7	0.5–0.6	0.6–0.7	0.6–0.7	0.5–0.6
	China	0.8	1.0	1.2	1.0	1.0	1.2	1.0
Fe, mg	WHO	5–10	5–10	10–24	14–28	5–10	9–18	5–9
	China		10–12	18	18	10–12	15	15
Zn, mg	WHO	10	15	15	15	15	15	15
Vitamin A, IU	WHO	400	575	725	750	575	725	750
	China	3,300	3,300	3,300	3,300	3,300	3,300	3,300
Vitamin D, IU	WHO	100	100	100	100	100	100	100
	China	400	400	400	400	400	400	400
Vitamin B_1, mg	China	1.2–1.4	1.2–1.4	1.5–1.6	1.5–1.6	1.2–1.4	1.5–1.6	1.5–1.6
Vitamin B_2, mg	China	1.2–1.4	1.2–1.4	1.5–1.6	1.5–1.6	1.2–1.4	1.5–1.6	1.5–1.6
Vitamin C, mg	WHO	20	20	30	30	20	30	30
	China	45–50	45–50	60	60	45–50	60	60
Folic acid, µg	WHO	100	100	200	200	100	200	200
Vitamin B_{12}, µg	WHO	1.5	2	2	2	2	2	2

were 350–1,100 kcal higher than the WHO RDAs for nonathletes (table 12). The distribution of energy sources was less than optimal, too much being derived from fat (40.3–44.6%) and too little from carbohydrate (41.8–47.5%). Further longitudinal studies are desirable, assessing the adequacy of nutrition against such criteria as growth in height and body mass, changes in the proportions of lean and fat mass and functional development.

Protein

Protein intakes averaged 1.8–2.4 g/kg with 10.9–16.1% of energy derived from protein. With few exceptions, the nitrogen balance of the athletes was positive. In hot environments, nitrogen loss in the sweat can lead to the development of a negative nitrogen balance; the nitrogen content of the sweat averaged 53.7 ± 14.3 mg/dl. When training in hot environments, athletes lost 7.1–10.9% of nitrogen by sweating [6]. We conclude that most athletes have an adequate protein intake, but that exceptions can arise in those undergoing intensive training, training in a hot climate, or having a limited overall intake of food energy.

Table 11. Tentative vitamin RDA for adolescent athletes [data taken from ref. 19]

	Age			
	7–9 years	10–12 years	13–15 years	16–19 years
Vitamin A, IU/day	4,000–5,000	4,000	5,000	5,000
Vitamin B_1, mg/day	1.2	1.2	1.2–1.4	1.2–1.4
Vitamin B_2, mg/day	1.6–1.7	2	2	2
Vitamin C, mg/day	85	85	80–90	80–90

Table 12. Patterns of food intake in Chinese male athletes [data taken from ref. 5]

Sports discipline	n	Age years	Protein g/kg	Fat g	CHO g/kg	Energy kcal	RDA
Gymnasts	6	10–11	2.1 (10.9%)	81 (44.6%)	181 (44.5%)	1,637 (100%)	2,200
Table tennis players	6	10–12	2.4 (12.2%)	119 (40.3%)	315 (47.5%)	2,654 (100%)	2,200
Wrestlers	12	15–18	2.3 (16.1%)	191 (42.4%)	439 (43.4%)	4,059 (100%)	2,850–3,000
Weight-lifters	14	15–18	1.8 (13.9%)	203 (44.3%)	430 (41.8%)	4,113 (100%)	2,850–3,000

Data in parentheses are protein, fat, and carbohydrate as percent of the total energy, respectively.

Minerals

Calcium insufficiency is quite common in Chinese amateur athletes. Not only is the total Ca intake low, but much is obtained from plant sources rather than milk. Thus, the absorption and utilization is likely to be poor, with a potential for adverse effects upon growth and development.

Trace Elements

The dietary intake of such elements as Fe, Zn and Cu met or exceeded the RDAs of the WHO, but deficiencies of Fe and Zn still arise (table 13). Hematological investigation of 449 child and adolescent amateur athletes found iron deficiency anemia in 22.5% of males and 57.9% of females [11, 12]. Causes for the iron deficiency seem complex and require further study. However, more than 90% of the iron was obtained

Table 13. Average daily intake of minerals observed in selected groups of Chinese amateur athletes [data taken from ref. 5]

Sports discipline	Ca mg	P mg	Fe mg	K mg	Na mg	Mg mg	Cl mg	Cu mg	Zn mg
Gymnasts	282	699	16.2	1,257	1,866	124	2,110	1.91	14.1
Table tennis players	669	1,383	25.4	2,392	3,945	301	4,757	2.78	26.3
Wrestlers	1,284	2,316	46.5	3,901	5,359	408	5,568	5.58	50.7
Weight-lifters	908	2,052	45.6	3,253	6,817	374	8,293	5.51	35.9

Table 14. Average daily intake of vitamins observed in selected categories of Chinese amateur athletes

Sports discipline	Vitamin A IU	Vitamin B_1 mg/1,000 kcal	Vitamin B_2 mg/1,000 kcal	Vitamin B_6 mg	Vitamin C mg
Gymnasts	1,421	0.66	0.29	7.93	24.3
Table tennis players	3,047	0.59	0.39	13.30	41.7
Wrestlers	3,772	0.33	0.39	13.55	69.8
Weight-lifters	7,222	0.59	0.34	17.69	52.4

from nonheme sources [5], so that poor absorption of iron may be an important contributory cause. Because of these findings, athletes now receive a daily prophylactic dose of 20 mg of iron; this appears to be effective.

A study of 75 adolescents found a poor Zn status in 12.3% and a poor Cu status in 35.1%. Serum Zn and Cu levels were 112 ± 22 and 99 ± 19 µg/dl, respectively, while hair Zn and Cu were 216 ± 45 and 14.1 ± 9.1 µg/dl. There is a need to explore whether these low values reflect poor absorption, increased utilisation, or increased daily losses.

Vitamins

Studies of vitamin intakes and loading tests show that about a third of athletes have deficient intakes of vitamins A, B_1 and B_2 (table 14). One cause of the deficiency was a lack of key foodstuffs in the diet (coarse grains, lean meat and milk or green vegetables). Athletes undergoing systematic training programmes have increased needs of vitamins B_1, B_2 and C [11, 12]. In general, adolescent athletes had good intakes of vitamins B_2 and C (1.7 and 35 mg/day, respectively) [5].

Conclusions

A comparison of athletes with average city students suggests that exercise accelerates growth and development. Exercise training also changes cardiac structure and morphology, with improvements of cardiorespiratory function, the proportion of body fat is decreased, and lean mass is increased. The energy requirements per unit of body mass are greater in children than in adults. Physical training increases the need for energy, proteins, iron, zinc, and vitamins. Surveys of athletes suggest that in general energy and protein intakes are adequate. The proportion of fat exceeds the RDA, but animal protein, milk calcium and heme iron are in short supply. Iron deficiency anemia seems common, and some athletes also show a low serum Zn and Cu. Some athletes have an inadequate intake of vitamins A, B_1 and B_2. Further study of interactions between nutrition, exercise and fitness are nevertheless needed, with particular reference to the growing child.

References

1. A Joint Research Report on Science and Technology: A study on the somatotype, functions and physique of Chinese children and adolescents. Beijing, Science and Technology References Publisher, 1982, pp 590–595, 638–643.
2. Chen JD, Chen ZM: A report on nutrition and health survey of children and adolescent athletes. A compilation of sports medicine materials, Beijing, Institute of Sports Medicine, Beijing Medical University, 1976, pp 169–174.
3. Chen JD, Yang ZY, Jiao Y, Bai RY, Chen ZM, Wu YZ: The nutrition and body composition of women gymnasts during weight control. J Sports Sci 1987;7:22–25.
4. Chen JD, Yang ZY, Jiao Y, Wu YZ, Chen ZM: A study on protein metabolism and requirement of athletes. J Sports Sci 1982;2:49–53.
5. Chen JD: Nutrition investigation of amateur children and adolescent athletes. Unpubl material, 1985–1986.
6. Chen JD, Yang ZY, Zhuo QL, Wu YZ, Chen ZM: Nutrition and metabolism of athletes training in hot environments. Chin J Sports Med 1987;6:65–69.
7. Chinese Nutrition Society: RDA of nutrition. Acta Nutr Sin 1990;11:93–96.
8. Gao YC, et al: Evaluation of body composition and equation for estimating the percentage of fat in college students. Chin J Sports Med 1984;3:76.
9. Health Institute of Chinese Medical Academy: Food Composition Table. People's Health Publisher, 1980, p 209.
10. Li ZY, Ding Z, Liu JB: The dynamic observation on cardiopulmonary function of adolescent middle and long distance runners during submaximum exercise. Chin J Sports Med 1984;3:225–230.
11. Li KJ: The study of RBC deformability in different physical loads and iron deficiency anemia; MSc thesis, Beijing Medical University, material unpubl, 1986.
12. Li YD: A primary study of vitamin B_1, B_2, and C requirements for juvenile athletes. Chin J Sports Med 1986;5:148–151.

13 Parizkova J: Growth, fitness and nutrition in preschool children. Prague, Charles University 1984, pp 9–14.
14 Parizkova J: Body composition and lipid metabolism in relation to nutrition and exercise; in Nutrition, Physical Fitness, and Health, Baltimore, University Park Press, 1987, pp 61–74.
15 Peltenburg AL: Biological maturation, body composition and growth of female gymnasts and control groups of school girls and girl swimmers aged 8–14 years: A cross-sectional survey of 1064 girls. Int J Sports Med 1984;1:1–56.
16 Qiao ZY, Wang AL: An echocardiographic analysis on the left ventricular function of 201 promising junior Chinese footballers. Chin J Sports Med 1984;3:218–222.
17 Qu MY, et al: Nutrition of athletes in 'Practical Sports Medicine'. People's Physical Education Publishing House, 1982, p 174.
18 Science and Education Department, All China Sports Federation: Selected Works on Researches in Sport Sciences and Technology, 1984, p 176.
19 Wu Han Medical College: Nutrition and Food Hygiene. Wu Han, People's Health Publishing House, 1981, p 21.
20 WHO Technical Report Series: Energy and Protein Requirements, Report of a joint FAO/WHO/UNU expert consultation – Rome 1981. Geneva, WHO, 1985.
21 Zhao YW, Chen JD, Wu YZ: Effects of endurance exercise on metabolism of zinc and copper. Chin J Sports Med 1990;9:10–14.
22 Zheng SQ: Measurements of body composition and equation for estimating the percentage of fat in college students. Chin J Sports Med 1984;3:76–80.
23 Zhou MY, Lu S: Age-related features of VO_2 max in male Chinese students from 13–23 years of age. Chin J Sports Med 1986;5:211–217.

Prof. Ji Di Chen, Research Division of Sports Nutrition and Biochemistry,
Institute of Sports Medicine, Medical University,
Beijing 100083 (People's Republic of China)

Social Epidemiology of Nutrition in the Ranga Reddy District of India and Its Implications for Human Resources Development

K. Satyanarayana[a], T. Prasanna Krishna[a], D. Banerji[b], B. S. Narasinga Rao[a]

[a]National Institute of Nutrition, Hyderabad; [b]Jawaharlal Nehru University, New Delhi, India

Introduction

This report describes the growth and development of rural Indian boys from the fifth year of life. The magnitude of undernutrition at the fifth year, the growth of body mass and height from the fifth year to adulthood (adolescent growth) and the quality of life attained by young adults is documented. Adolescent growth is examined as the outcome of such independent variables as the socioeconomic status of the family, childhood undernutrition and childhood occupational stress. The current theory of undernourished ('small but healthy') children is also evaluated.

Data on the heights and weights of 900 adolescents were collected over a period of 7 years from 1976 to 1983. An earlier study conducted by the National Institute of Nutrition in Hyderabad had surveyed those currently young adults as very young children [1]. Data were available from growth records maintained between 1965 and 1969.

A combination of the data collected between 1976 and 1983 with results from the earlier study [1] has enlarged the scope of our findings, providing a total of 18 years study between 1965 and 1983. Children who were 1–5 years old in 1965 were aged between 19 and 23 years in 1983. These figures allow us to assess long-term growth, over a 15-year period beginning from the fifth year of life, and ending with the cessation of linear growth. The extent of undernutrition at the fifth year, growth from the fifth year to adulthood, literacy levels, social placement and the income earned as adults have been examined in relation to caste, economic class, poverty status and literacy status of the individual families.

Methods

Undernutrition at Fifth Year

The nutritional status of about 1,000 boys was followed over a 4-year period (1965–1969), with assessment at half-yearly intervals. Twenty-three villages of the Hyderabad East District, now called Ranga Reddy District, were studied. Height and body mass were available on all boys from the fifth year of life. Both international [2] and national [3–5] reference standards were used for comparison. The extent of undernutrition was examined in similar land-ownership classes among four caste groups. Some of the villages were in the process of semiurban development by 1977.

Adolescent Growth

About 900 boys from the 23 study villages were contacted in 1976. Some 10 villages had definitely become semiurbanised by 1983. Heights and body masses were measured from 1976 to 1983, in eight annual surveys, each taking about 4 months to complete (December to April). Measurements in a given village were carried out during the same period in each succeeding year. Repeated visits were made to find missing subjects. The survey team resided in a central village, to be close to the people and thus to obtain a better rapport with the subjects. Two special surveys accumulated socioeconomic information. The first survey (in 1977) collected demographic information on the family and its members. The second survey (in 1983) collected information on social placement and life patterns.

Socioeconomic Data Base

During 1977, data were collected on caste, family size and composition, educational levels and all-source family income. Family land-ownership data were obtained for 1969 and for 1977, the 1969 data being used to examine the relationship to undernutrition at the fifth year.

Family resources of (1) income from all sources in 1977; (2) literacy levels of three family members, and (3) occupation of the head of the household were each assigned arbitrary scores [6]. The total of the three scores was used to separate families into two groups, designated as (1) 'poor' (with 4–9 family resource points), and (2) 'well-to-do' (with 10–30 family resource points). The per capita income of the poor group was always less than Rs. 60/– per month in 1977, an amount judged as inadequate to buy a balanced diet. The majority of families falling into this group had family per capita incomes of Rs. 25/- to Rs. 45/– per month. The social hierarchy (caste) position of the families was grouped into four classes as suggested by sociologists [6].

Data on employment patterns, earned income, literacy levels and life-styles were examined in 1983. The family resource score of 1977 was taken as the baseline to examine growth, educational opportunities and social placement attained by 1983.

Methodology of Growth Assessment

A growth in stature of less than 1 cm over a full year was taken as the point of adulthood. Growth data obtained between 1981 and 1983 showed that 408 men aged 19–23 years had completed their linear growth. The adult height and body mass were compared to international [2, 7] and national [4] reference values. The total growth of height and body mass from the 5th year to adulthood was obtained by subtraction of the 1965–1969 data from that collected between 1981 and 1983. This growth was compared with that of longitudinal data for boys in Boston, Mass. [2].

Concept of Social Epidemiology of Nutrition and Quality of Life

Social, political and economic forces play a large role in determining the meaning of ill health [8, 10], but the social and economic dimensions of childhood undernutrition and adolescent growth have not previously received adequate study.

Using epidemiology as the basic tool, social dimensions have been built into the framework of our present analysis [10]. Previous sociological surveys have often been criticized for making qualitative statements, with limited quantitative analysis of data. A serious attempt has been made here to analyse the data in order to avoid this criticism. Multiple correlation analysis has been undertaken, with growth as the dependent variable. Independent variables have included (1) family variables, like caste status and family resource score; (2) exploitation of the child for wage-labour, and (3) the impact of childhood growth retardation upon adolescent growth. Partial correlation analyses have been carried out. Other analyses have examined the number of years of education completed and the current income in 1983.

Results

Extent of 'Undernutrition' at the Fifth Year

The combined prevalence of severe and moderate growth retardation assesses the 'quantum of undernutrition' in a community. Severe (height for age $<$ mean-4 SD [2]) and moderate (mean-3 SD to mean-4 SD) growth retardation in children was observed in about 20 and 28% of children, respectively. Severely undernourished children had a height deficit of 19 cm and a body mass deficit of 7 kg relative to 5-year-old children from well-to-do Indian families [3]. A normal nutritional status (height between mean and mean-2 SD [2]) was noticed in 21% of the children and the remaining 31% showed mild growth retardation.

Undernutrition was observed in only 18% of children from landlord families. In families without land, this proportion increased to 57%. On the other hand, about 50% of children from landlord families enjoyed a normal nutritional status, while only 13% of children from land-less families achieved this status in their fifth year of life. Some of the study villages became peri-urban and semi-urban by 1977, and by 1983 urbanization was much greater; however, no attempt was made to separate data from semi-urban villages. The family resource score provided a built-in correction to account for the real value of money in rural and semi-urban situations.

Family land-ownership data revealed that Forward Caste group families dominated the landlord class. Backward Caste group families dominated the medium level land-holding class. However, this particular group had the maximum family size. Scheduled Caste group families either owned smaller holdings or had no land at all.

Families with no land-holdings and families with limited land-holdings had the highest quantum of undernutrition and the lowest percentages of normally nourished children. Large family size and a higher level of parental deaths further aggravated this situation.

Families without any land from 1969 onwards and who continued to have only 4–9 points of family resources till 1977 were taken as abjectly poor families. Altogether, there were 390 families without any land in the total sample of 894 families. Out of these, 49 families moved up the family resource scale by the year 1977, but the remaining 341 families were classified as abjectly poor. The nutritional status of children from abjectly poor families was similar to that of the original 390 land-less families, from which this group was separated. Only about 15% of abjectly poor children enjoyed a normal nutritional status, and about 57% of children from this most disadvantaged subgroup suffered from undernutrition. It was concluded that during the 1960s, the period when the study area remained rural, a simple land-ownership classification of caste groups was equally good. At this stage, there was no need for complex socioeconomic scores to understand the pattern of concentration of economic resources. On both classifications, it was very clear that the Forward Caste group had a very high concentration of resources for half of its members. Hardly 8% of the Scheduled Caste group families had medium-sized land holdings. Poverty was seen in all caste groups (23 to 45%), but the proportion was maximum (45%) in the Scheduled Caste group.

Childhood Occupational Stress and Social Class

'Child labourers' were identified as those who had undertaken a minimum of 4 years of agricultural wage-earning work before reaching the age of 14 years [11]. About 20% of the children had worked as child-labourers. Some of them had started working as young as their sixth year of life. About 43% of children from the Scheduled Caste group had worked as child-labourers, but this was true of only 4% of children from the Forward Caste group. More than half of the children from the Scheduled Caste families had never attended school in their lives. In contrast, three-quarters of children from the Forward Caste families were still attending school by their 14th year. Practically all child-labourers (98%) were drawn from poor families, irrespective of their caste status. Poverty, landlessness, illiteracy, dependency on other families and family indebtedness have all perpetuated child labour in this society.

Adult Nutritional Status of Undernourished Children

The final adult heights and body masses were significantly lower in severely undernourished children (fifth year category) when compared to other groups [11]. Very short children had a height deficit of 10 cm and a body mass deficit of 7.5 kg as young adults. These deficits reached 16–26 kg for body mass and 14–21 cm for height, when compared with well-to-do Indian [4] or Western adults [2, 7, 11]. A major proportion of severely undernourished children had to work as child-labourers; most of them were drawn from abjectly poor families and most often from families with low social caste. No supporting evidence was found for the 'small but healthy' hypothesis [12]. Undernutrition, child labour, illiteracy, poverty and the dependent status of families usually went together, and the undernutrition seen in the children was an indication of social deprivation. Attempts to justify severe or moderate undernutrition would seem to perpetuate social injustice and unequal distribution of available resources. The 'small but healthy' hypothesis indirectly supports social exploitation in the guise of technical jargon and was seriously criticized [13].

Multiple Correlation Analyses

Correlation coefficients indicate that childhood and adolescent growth alone could explain some 81 and 30% of the variance in adult body mass and height, respectively. Child labour significantly reduced growth, even after adjustment of data for family caste status and family resources. On the other hand, the better growth of higher caste group adolescents was realized only by virtue of greater family resources. Caste status alone, unless accompanied by greater resources, had no influence on growth. The greater growth of adolescents from higher caste groups is due to their families' sound economic position, and to the absence of exploitative occupational stresses.

Literacy Levels

All young men from well-to-do families had completed elementary school education. When family resources were adequate, the percentages of literates among fathers (46–47%) and sons (80–94%) in the two extreme caste groups were nearly equal. Family resources during 1977 explained the educational opportunities available to young men over the next 6 years. However, a very small proportion (9%) of Scheduled Caste group families had family resources which were worth 10 points or more. A nearly similar proportion of young men from such families with adequate resources had reached junior college (10 + 2) as compared to other higher caste groups. They had also tended to obtain well-paid factory jobs.

Unlike the above picture, there were substantial differences of literacy levels for fathers and sons of poor families from extreme caste groups, two-fold for elementary education, but ten-fold for junior college (10 + 2) education. Poor Scheduled Caste families depended on other caste group families for daily sustenance and loans in emergencies. This dependency, the illiteracy of the fathers and above all the lack of control over land and other means of production rendered them helpless. In order to survive, they had to send their young children for wage-labour. Lack of family resources created a vicious cycle of continuing illiteracy and exploitation. Partial correlation analysis revealed independent influences of both caste status and family resources on the number of years of education. For this parameter, caste hierarchy status had an independent influence, even after adjusting data for the influence of family resources. Scheduled Caste children at younger ages were thus sent to child-labour instead of to school, even at comparable levels of poverty or family resources. It is concluded that higher caste groups exploit children from poor families, particularly if they belong to lower caste groups.

Life Patterns as Young Men

The most exploited and under-paid group were the child-labourers, who continued to work on annual contract ('Jeetam') for landlord families. They were paid one-fourth of the wages earned by factory workers. Even normal agricultural labour paid only about 58% of semi-urban factory wages. At the age of 20 years, a poor village worker earned Rs. 60/- to Rs. 150/- per month (1 US$ = Rs.10/-) in 1983. Out of the 52 young men who were working for wages in the agricultural sector, 43 were child-labourers in early life and the majority belonged to poor Scheduled Caste families. The tendency to movement of young men away from the villages had definite caste and poverty dimensions. One had to be equipped with elementary literacy, a knowledge of the outside world and an initial capacity of the family to sustain a period of absence, in order to make contacts with the outside world. Mobility from the villages was associated with a two- to fourfold increase in the wages earned. There were only limited chances for those from poorer families and low social classes to move, but if they succeeded in moving, they earned equal amounts, irrespective of caste and economic background. However, the opportunities for children to pursue higher education were substantially lower for children from poor families, particularly if they were from the poor Scheduled Caste. About 50% of those from the well-to-do Forward Caste group and 30% of those from the well-to-do Scheduled Caste group were attending educational institutions in 1983, but only 8% of those from the poor Scheduled Caste group could continue their studies. Caste status as well as family resources influenced the ability to continue education.

Conclusions

There were remarkable differences between extreme caste groups with respect to both land-ownership in 1969 and total family resources in 1977. Family resources were concentrated among a few families from the Forward Caste group and to some extent from the Backward Caste group. Undernutrition was rare among children from landlord families and the majority of their children enjoyed a normal childhood nutritional status. Almost all such children completed their school education and 40–50% were continuing their studies in 1983, having reached an age of 19–23 years. Growth was greatest in this group, and they achieved nearly optimal heights and body masses as young adults.

Scheduled Caste group families had very limited land ownership and family resources in 1977. The few well-to-do Schedule Caste families behaved differently and were similar to other well-to-do families of higher caste groups. Poor Scheduled Caste families were the most deprived group, usually illiterate and dependent on other caste groups for day-to-day existence. Exploitation of these families created conditions and economic obligations which led to child-wage labour. In consequence, the child's learning was kept to a minimum by the land-owning class. Child labourers accounted for much of the undernutrition seen at the fifth year of life. The majority of Scheduled Caste group children never went to school, and 43% worked for wages. More than half of them suffered from growth retardation by the fifth year of life and developed additional growth deficits during adolescence. Most of these poor Scheduled Caste children were given poor wages for agricultural work, even as young adults.

All of these problems are by-products of poverty, and lasting improvement can only be obtained by improving purchasing capacity through proper employment-generating programmes.

It is plainly undesirable to examine only the clinical nutritional status or the physical anthropometry of children without considering their social circumstances. It is unscientific to conclude that undernourished ('small') children are healthy [13], and it seems a disservice to humanity to propose less than optimal growth standards for certain countries in order to justify undernutrition and label it as social undernutrition.

References

1 Swaminathan MC, Susheela TP, Thimmayamma BVS: Field prophylactic trial with a single annual oral massive dose of vitamin A. Am J Clin Nutr 1970;23:119–122.
2 Reed RB, Stuart HC: Patterns of growth in height and weight from birth to eighteen years of age. Pediatrics 1959;24:904–921.
3 Hanumantha Rao D, Satyanarayana K, Gowrinath Sastry J: Growth pattern of well-to-do Hyderabad preschool children. Ind J Med Res 1976;64:629–638.
4 Hanumantha Rao D, Gowrinath Sastry J: Growth pattern of well-to-do Indian adolescents and young adults. Ind J Med Res 1977;66:950–956.
5 Vijayaraghavan K, Darshan Singh, Swaminathan MC: Heights and weights of well-nourished Indian school children. Ind J Med Res 1971;59:648–654.
6 Narayana Rao S: The socio-economic status rating scales (SESR scale). Ind J Soc Sci 1973;Sept:206–219.

7   Eveleth PB, Tanner JM: Europeans in Europe and European descendants in Australia, Africa and the Americas; in: Worldwide Variation in Human Growth. London, Cambridge University Press, 1976, IBP vol 8, pp 15, 51.
8   Banerji D: Epidemiological issues in nutrition. Ind J Nutr Dietet 1979;16:189–194.
9   Banerji D: Poverty Class and Health Culture in India. New Delhi, Prachi Prakashan, 1982, vol 1.
10 Suchman EA: Sociology and the Field of Public Health. New York, Russel Sage Foundation, 1963.
11 Satyanarayana K, Prasanna Krishna T, Narasinga Rao BS: Effect of early childhood undernutrition and child labour on growth and adult nutritional status of rural Indian boys around Hyderabad. Hum Nutr Clin Nutr 1985;40C:131–139.
12 Seckler D: 'Small but healthy': A basic hypothesis in the theory, measurement and policy of malnutrition; in Sukhatme PV (ed): Newer Concepts in Nutrition and Their Implications for Policy. Pune, Maharashtra Association for Cultivation of Science Research Institute, 1982, pp 127–137.
13 Gopalan C: Small is healthy? For the poor, not for the rich. Nutr Found Ind Bull 1983;Oct:1–5.

K. Satyanarayana, MD, National Institute of Nutrition,
Indian Council of Medical Research, Jami-Osmania, Hyderabad–500007 (India)

Growth, Maturation, Body Composition and Maximal Aerobic Power of Nutritionally Normal and Marginally Malnourished School-Aged Colombian Children[1]

G.B. Spurr[a], *M. Barac-Nieto*[b], *J.C. Reina*[b]

[a]Department of Physiology, Medical College of Wisconsin and Research Service, VA Medical Center, Milwaukee, Wisc., USA; [b]Departments of Physiological Sciences and Pediatrics, Universidad del Valle, Cali, Colombia

During 1978–1982, we carried out an extensive survey of the growth and physical fitness of over 1,000 school-aged boys in relation to their nutritional and socioeconomic status in Cali, Colombia, and its rural environs [1, 2]. This was done to characterize the influence of chronic, marginal malnutrition on the growth of aerobic power in these children and to relate it to the effects that might be expected on the work capacity of the subjects when they achieve adulthood. The significance of this is related to the fact that in developing countries, a high percentage of adult males are working at physically taxing jobs. A recent estimate of nine developing countries located in the tropics showed that an average of 60% of the adult male work force was engaged in moderate-to-heavy physical work [3].

Then, in the years 1982–1987, we also measured the physical work capacity of 218 boys and 136 girls as part of a study on the effect of nutritional status on daily energy expenditure and to obtain a comparison of the maximum oxygen consumption ($\dot{V}_{O_2}$ max) of school-aged boys and girls growing up in the disadvantaged economic circumstances of a developing country [4]. These two studies will be referred to as study I and study II, respectively.

Colombia is a country considered to be at a middle level of development. Cali is an industrial city of about 1.7 million inhabitants, the third

---

[1] These studies were supported by NIH Grant HD10814, Nestlé Nutrition, the United Nations University, the Research Service, VA Medical Center, and the Fundación para la Educación Superior, Colombia.

largest in Colombia, located 3° 22′ north of the equator at an altitude of 976 meters. In common with other Latin American cities, during the past 25–30 years it has undergone rapid growth due, in part, to an influx of population from rural areas. It enjoys a year-around average temperature of some 24 °C (high 29, low 18 °C) which varies little throughout the year, so that wide seasonal differences in ambient temperature were not a factor in the studies to be described. There are two rainy seasons (March to June and October to December) during which average monthly rainfall may reach a maximum of 18 cm, while maximum rainfall during the dry seasons is 6–7 cm/month.

Methodology

The methods used have been described previously in more detail than is necessary here [1, 2, 4]. Briefly, studies I and II were conducted in an air-conditioned mobile laboratory which allowed for moving into the urban neighborhoods or rural areas where the subjects were living. Contacts were made with the subjects through the public schools attended by most of the children, or the private schools which the economically advantaged children of study I attended.

Subject Groups
Birth certificates were required of all subjects in order to be included in the study.

Study I. On the basis of preliminary weights and heights the subjects were divided into three nutritional groups using the Colombian norms established by Rueda-Williamson et al. [5]. Those characterized as nutritionally normal (control) were within 95–110% of these norms of weight-for-age and weight-for-height. The low weight-for-age group (Wt-Age) were those who were <95% of weight-for-age but 95–110% of weight-for-height and were considered to be nutritionally normal at the time of the study but as having a history of malnutrition sometime in the past. Finally, the low weight-for-height (Wt-Ht) were those children who were <95% of the norm in both weight-for-age and weight-for-height and were considered undernourished at the time of the study.

The children were also separated into three socioeconomic status (SES) groups: upper (U) urban, lower (L) urban, and lower rural on the basis of where they lived and family income. There were no upper socioeconomic groups living in a rural area since all such families deriving their income from agricultural pursuits live in Cali to take advantage of better schooling for their children. There are four groups of urban children (USES controls, LSES controls, L Wt-Age and L Wt-Ht) and three nutritional groups of rural LSES children. The subjects were also separated into five age groups of 2-year intervals between 6 and 16 years of age. Therefore, there were 35 groups of boys with n values varying between 13 and 57 in each group. The $\dot{V}_{O_2}$ max was measured in a total of 1,013 boys [2].

Study II. In this second phase, only three age groups of boys and girls were studied: 6–8, 10–12 and 14–16 years old. Also, the children were all LSES, living in the economically deprived urban areas of Cali and were classified as nutritionally normal controls and marginally undernourished (L Wt-Ht, i.e. weight-for-age and for-height both <95% of predicted [5]). Thus, there were 12 groups of children with n values of 15–54 each. They were selected in the same way as the boys in study I.

Anthropometry and Body Composition

Weights were obtained on a Homs beam balance (± 25 g) while wearing shorts and, in the case of girls, a light blouse. Heights were measured with a wall stadiometer. Triceps and subscapular skinfolds were measured in triplicate by a Lange caliper and mid upper-arm circumference by flexible tape. In 53% of the subjects in study I, it was possible to remeasure weights and heights 6–12 months after the first set of measurements and so to calculate growth velocity.

We have shown by measurements of total body water and derived estimates of body fat and lean body mass (LBM) that the equations developed by Pařízková [6] for European children are applicable to our Colombian boys, whether nutritionally normal or undernourished [7]. Using these equations and measurements of triceps and subscapular skinfolds, body density was predicted, and body fat and lean body mass obtained from the Brožek formula [8] in the usual manner.

Maturation

The sexual maturation of the children was judged by one member of the team (J.C.R.) by assigning Tanner Scores 1–5 as described by Tanner [9].

Maximum Oxygen Intake ($\dot{V}_{O_2}$ max)

The $\dot{V}_{O_2}$ max was measured directly on a motor-driven treadmill using a modified Balke and Ware [10] procedure which, after a 3-min warm-up at 3.5 mph at a 5% grade, continued at 3, 3.5 or 4.2 mph (depending on the heart rate response during warm-up) and 2.5% grade increases each 2-min until exhaustion. Details have been presented previously [2]. The children all attended a preliminary training session on a separate day and were tested in small groups to provide competition and mutual stimulation. The results were that more than 80% of the boys and 72% of the girls achieved maximum values as judged by a number of criteria [2, 4]. Only data from children who achieved maximum effort are reported in the maximum exercise related variables.

Blood Chemistries

A finger-tip blood sample was used to measure blood hemoglobin by the cyanmethyemoglobin method.

Results

Anthropometry and Body Composition

Study I. Average weights and heights of the seven socioeconomic and nutritional groups in study I are plotted as a function of mean group age in figure 1 to compare with the percentiles of the US National Center for Health Statistics (NCHS) data [11]. The LSES, nutritionally normal urban and rural boys were significantly shorter than the USES boys, but not different from each other. Despite the fact that in the statistical manipulation, race was held constant, the difference is no doubt due to the difference in racial composition of USES (predominantly caucasian) and LSES (predominantly mestizo). When the analysis is done only on mestizo boys,

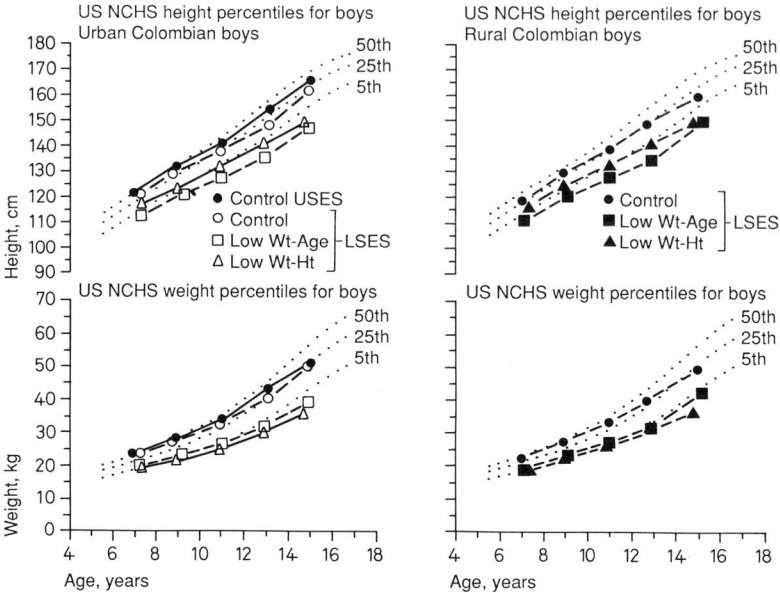

Fig. 1. Average values of height and weight of boys in study I plotted as a function of average group ages on the US National Center for Health Statistics (NCHS) percentiles. The SEs varied between 0.6 and 1.6 cm for height and 0.2 to 0.9 kg for weight (data from Spurr et al. [1]).

all differences between the nutritionally normal USES, and urban and rural LSES boys disappear. There are no differences in weight between these three groups.

It is clear from figure 1 that the boys with a history of undernutrition (L Wt-Age) and those considered to be undernourished at the time of the study (L Wt-Ht) were undergoing a process of stunting in growth such that both height and weight fell on or below the 5th percentile of the US NCHS data. A 3-way ANOVA of only the two undernourished groups by age and urban-rural site demonstrated significantly lower weights of the L Wt-Ht compared to the L Wt-Age boys ($p < 0.01$) but the former were significantly taller than the latter ($p < 0.001$). There was no statistically significant urban-rural difference in height but the rural boys weighed significantly less than the urban children ($p < 0.01$).

Body composition variables of the seven socioeconomic and nutritional groups are shown in figure 2 together with the results of the 3-way ANOVA for urban-rural effects and 2-way ANOVA for age (A) and nutritional group (NG) effects in the urban and rural boys separately.

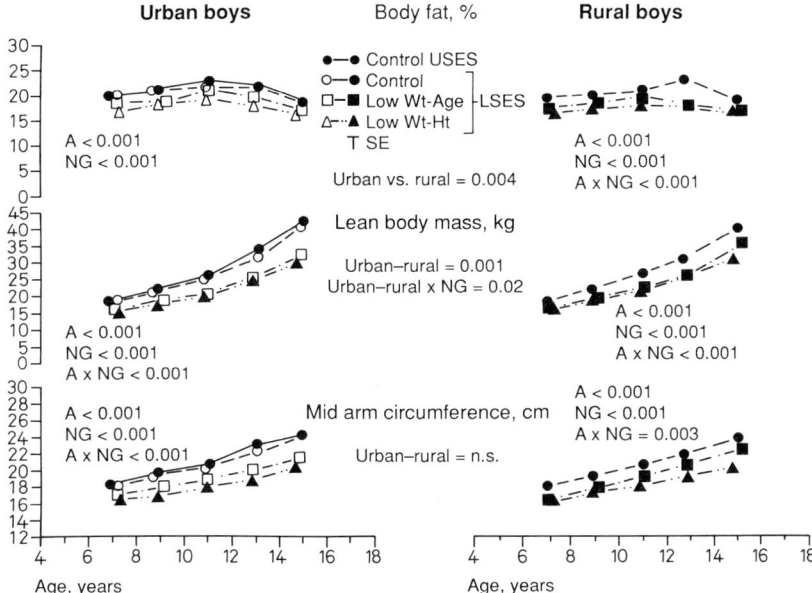

Fig. 2. Body composition of boys in figure 1 with results of 3-way ANOVA for Urban-Rural differences and 2-way ANOVA for age (A) and nutritional group (NG) effects. Symbols cover SE bars. (data from Spurr et al. [1]).

There were significant age and nutritional group effects in all 2-way ANOVAs and the rural boys had smaller LBM and body fat values than urban subjects. These differences were due entirely to differences in the undernourished boys since there were no significant differences in LBM and body fat of the urban and rural normal boys. There were no statistically significant differences in mid-arm circumference values between the boys living in the two locations (fig. 2).

Study II. In terms of weight and height, the boys and girls of the second study exhibited similar changes to those seen in the boys of study I except that the older girls seem to be less stunted than the boys (fig. 3). Body composition also showed significant age and nutritional group effects with the girls having higher values for body fat and lower LBMs than boys without a significant difference in the sexes in the mid-arm circumference (fig. 4).

Growth Velocity

The growth velocities calculated from the second weighing of 53% of the subjects in study I, 6–12 months after the first, are plotted in figure 5.

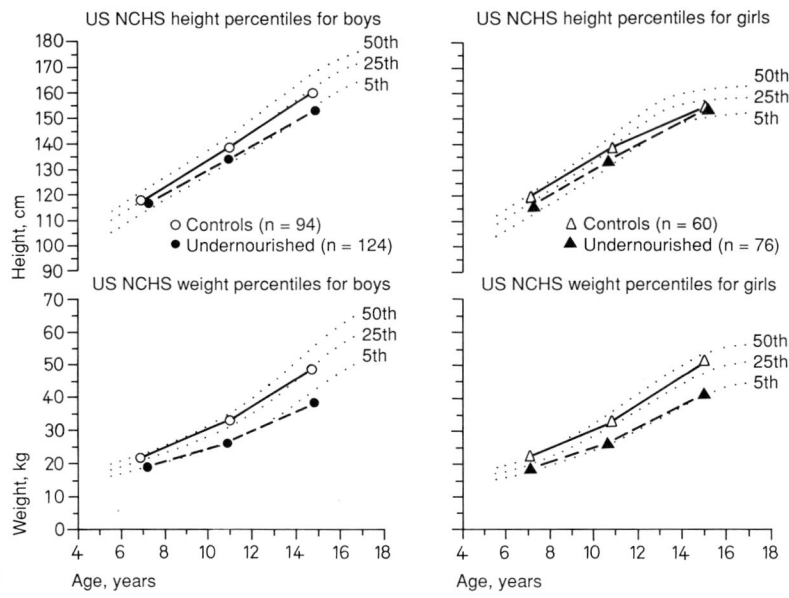

3

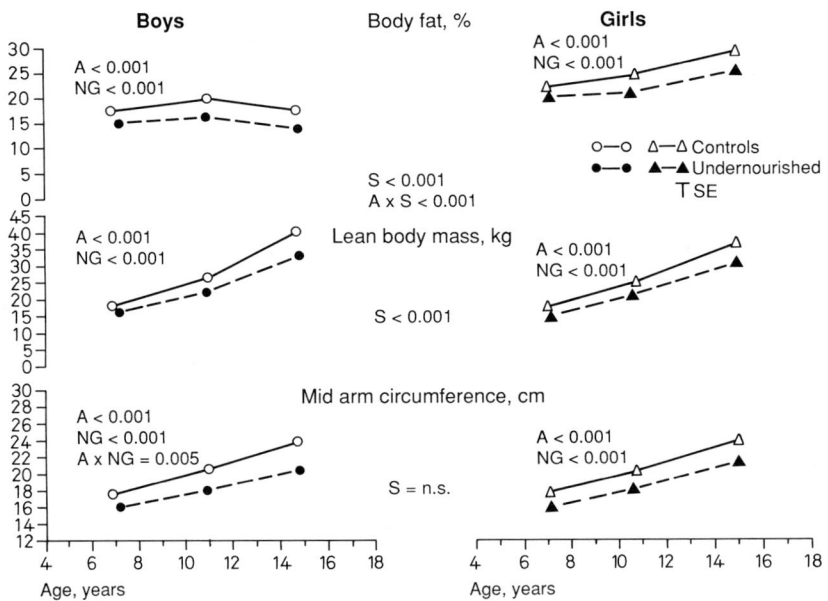

4

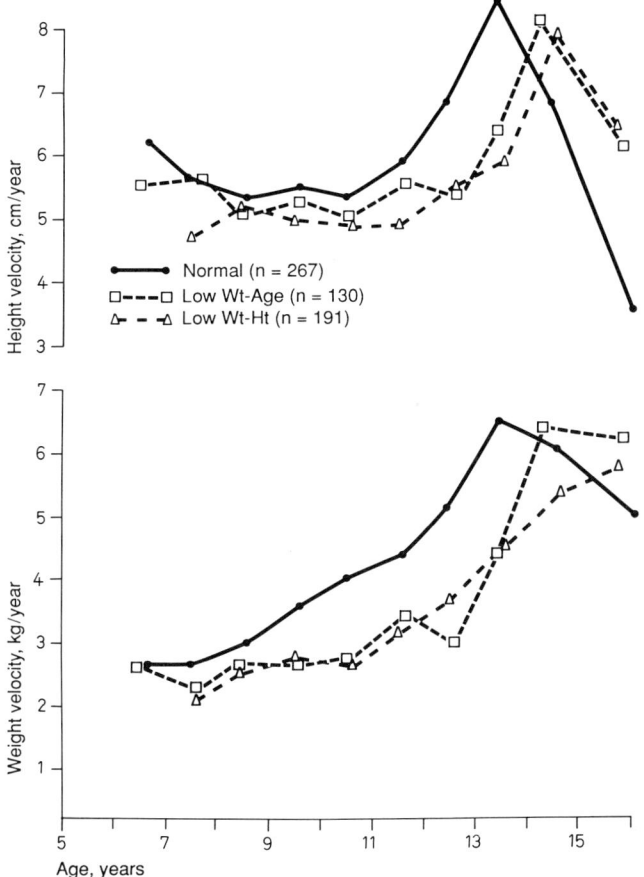

Fig. 5. Average growth velocities of nutritionally normal, low Wt-Age and low Wt-Ht boys plotted at the average age midway between two measurements obtained 6–12 months apart. SEs varied between 0.1 to 0.9 cm/year for height velocity and 0.1 and 1.0 kg/year for weight velocity (reprinted with permission from Spurr et al. [1]).

Fig. 3. Average values of height and weight of the children in study II plotted as a function of average group ages on the US National Center for Health Statistics (NCHS) percentiles (data from Spurr and Reina [4]).

Fig. 4. Body composition of children in figure 3 with results of 3-way ANOVA for sex (S) and 2-way ANOVA for age (A) and nutritional group (NG) effects. Symbols cover SE bars (data from Spurr and Reina [4]).

The loss of subjects in the second weighing was a random one since the racial composition of the three socioeconomic groups did not change and there were no significant differences in the mean initial heights and weights of the children who were remeasured compared to the total group. Since there were no detectable socioeconomic differences, the subjects were combined into the three nutritional groups.

The height velocity curves of the two nutritionally deprived groups are displaced to the right of that in normal boys, but the differences did not achieve statistical significance ($F = 2.21$; $p = 0.11$). The weight velocity curves of the two deprived groups were lower and also displaced to the right of the control children, and the difference was statistically significant ($F = 12.88$; $p < 0.001$). There were no significant differences between the growth velocity curves of the two nutritionally deprived groups. There was a statistically significant and progressive delay in the age at which peak velocity was achieved in normal, L Wt-Age and L Wt-Ht boys, respectively, for both height and weight [1]. So there is no doubt that the boys who were classified as nutritionally deprived were affected in their growth rates.

Sexual Maturation

The frequencies of the 14- to 16-year-old children in study II, with Tanner scores of 1–5, are shown in table 1 [4]. The χ^2 analyses demonstrate statistically significant retardation of sexual maturation in both boys and girls. A similar finding was observed in the 12–13.9 and 14–16-year-old boys who were undernourished in study I [1].

Table 1. Frequencies of 14- to 16-year-old boys and girls with indicated Tanner score in study II

Nutritional groups	Tanner score					x^2 analysis
	1	2	3	4	5	
Boys						
Control	2	0	10	9	11	$\chi^2 = 21.18$
Undernourished	17	4	5	7	4	$p < 0.001$
Girls						
Control	0	0	0	9	12	$\chi^2 = 6.61$
Undernourished	0	0	5	7	7	$p < 0.05$

Physical Condition and Aerobic Power

In the subjects in whom the $\dot{V}_{O_2}$ max was measured in both studies I and II, the group averages for maximum heart rate were in excess of 200 [2, 4] with the girls having values which were slightly, but significantly higher than boys [4]. The individual group means and SD's of heart rate, pulmonary ventilation, and $\dot{V}_{O_2}$ max in terms of liters/min, ml/kg · min body weight and ml/kg · min LBM can be found in the original publications [2, 4].

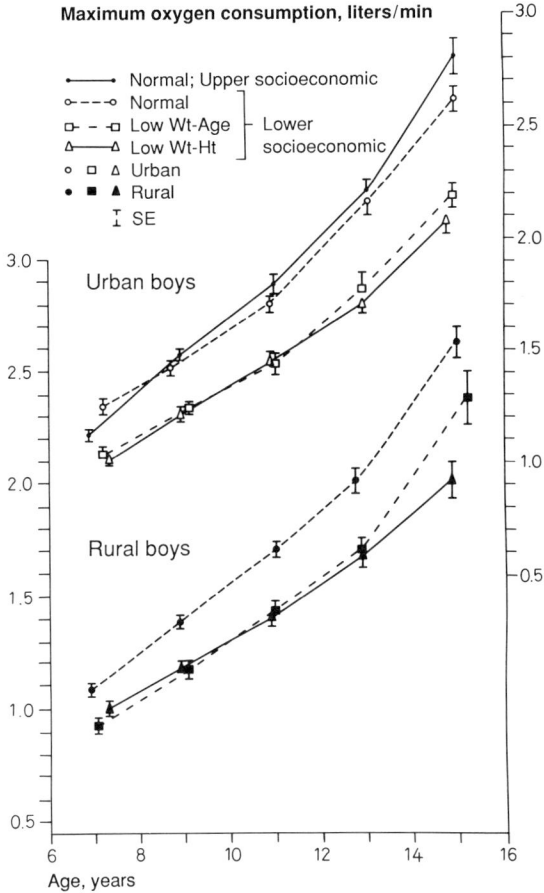

Fig. 6. $\dot{V}_{O_2}$ max (liters/min) of nutritionally normal, low Wt-Age and low Wt-Ht urban (right scale) and rural (left scale) boys in study I. Plotted as a function of average group ages (reprinted with permission from Spurr et al. [2]).

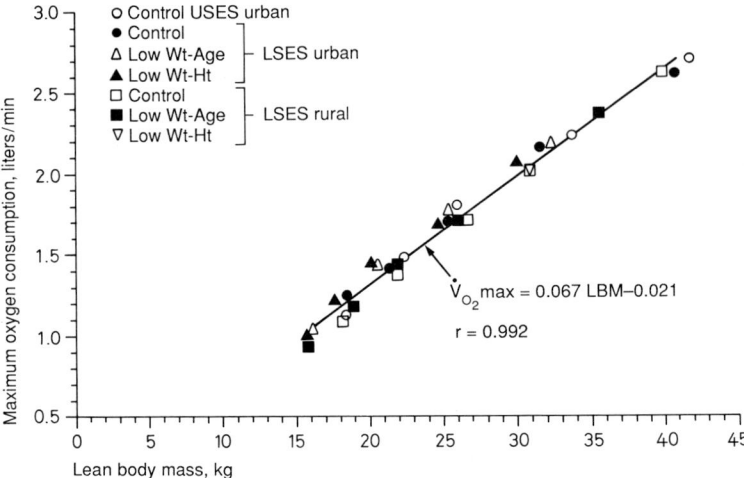

Fig. 7. Averages of $\dot{V}_{O_2}$ max of 35 groups of boys in study I plotted on group averages of lean body mass (data from Spurr et al. [2]).

Since the differences in racial mixture of the upper and lower socioeconomic groups [1] are a possible confounding variable in the present data, we analyzed the results obtained in the control groups of boys by race (106 white, 217 mestizo, 70 black) and reported that there were no significant differences in $\dot{V}_{O_2}$ max (liters/min or ml/kg · min body weight), or in the regression of $\dot{V}_{O_2}$ max (liters/min) on body weight between the three races [12]. Consequently, in what follows, race is not a factor in the differences due to nutritional status.

The mean values of $\dot{V}_{O_2}$ max (liters/min) in study I are plotted in figure 6 as a function of average group age. There were no statistically significant differences between the two nutritionally normal groups nor between the two nutritionally deprived groups of urban boys. But the latter two groups (L Wt-Age, L Wt-Ht) were significantly lower than control boys throughout the age range studied ($F = 69.38$; $p < 0.001$). A similar pattern is seen in the rural boys; L Wt-Age and L Wt-Ht boys and $\dot{V}_{O_2}$ max (liters/min) values which are not significantly different from each other but which are significantly lower ($F = 45.72$; $p < 0.001$) than their nutritionally normal counterparts. Also, the values for rural boys were significantly lower than those for urban children ($F = 5.19$; $p = 0.023$).

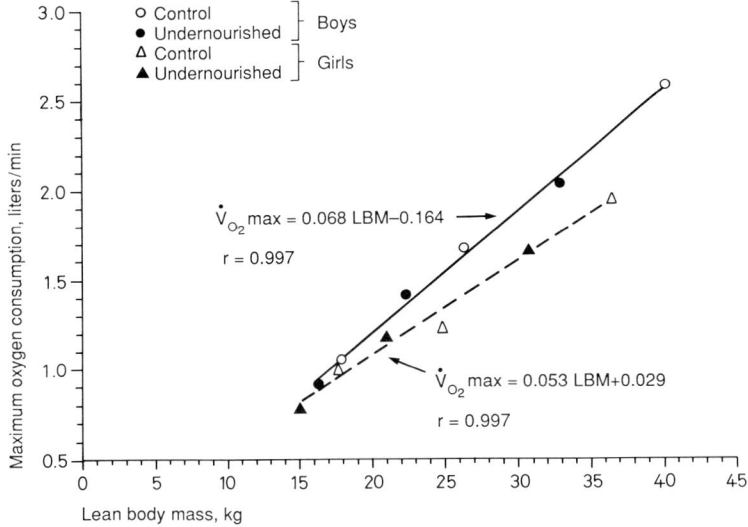

Fig. 8. Averages of $\dot{V}_{O_2}$ max of the 12 groups of boys and girls in study II plotted on group averages of lean body mass (data from Spurr and Reina [4]).

A similar pattern was observed for the boys and girls in study II, i.e. significant increase with age and lower values in undernourished than control subjects. There were also lower values in girls than in boys [4]. We have suggested that the reduced $\dot{V}_{O_2}$ max in marginally undernourished children is largely a result of the differences in their body size. This is born out by plotting the data from the two studies as liters/min on LBM as shown in figures 7 and 8. In study I, the average $\dot{V}_{O_2}$ max in liters/min of all 35 groups fall on the same straight line when body size is expressed as LBM. There were no statistically significant differences in the slopes or intercepts of the seven lines formed from the different nutritional and socioeconomic groups so they were combined in the regression analysis presented in figure 7. When the boys and girls of study II are plotted in the same way (fig. 8) there is a statistically significant difference ($p < 0.001$) between the sexes in the slopes of the two lines so formed [4], which means that the girls are in poorer physical condition than the boys. The linear relationship of the boys in study II (fig. 8), who were different subjects from those in study I (Fig. 7), was similar to that obtained in the latter.

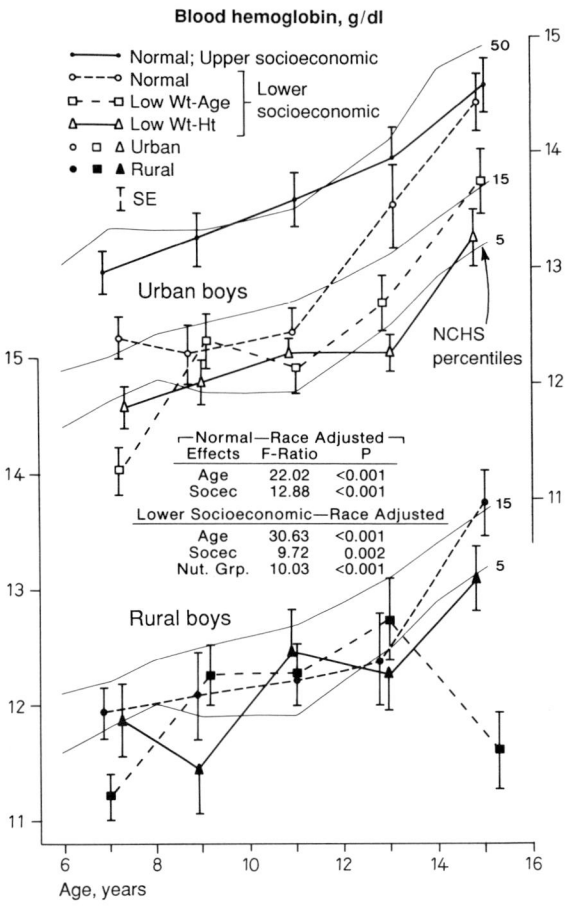

Fig. 9. Blood hemoglobin concentrations of urban (right scale, upper panel) and rural (left scale, lower panel) by nutritional and socioeconomic groups as a function of average group ages with results of 2- and 3-way ANOVA. Percentile grids are from US NCHS (Garn et al. [13]; reprinted with permission from Spurr et al. [2]).

Blood Hemoglobin

Study I. In figure 9 the mean values for blood hemoglobin are plotted on the US NCHS percentile grids [13] and show the statistical analysis of the data. The 2- and 3-way ANOVA were adjusted for race since it is known that blacks have lower blood hemoglobin values than whites [14]. There was a progressive and statistically significant increase with age ($p < 0.001$). Furthermore, there were significant socioeconomic differences

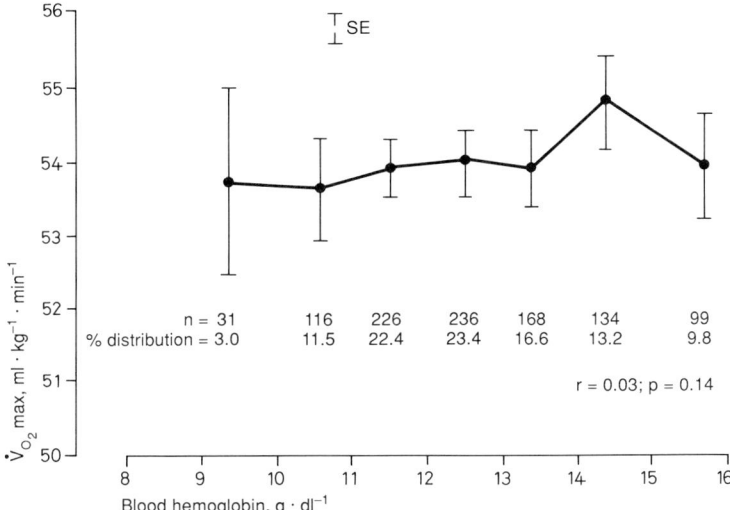

Fig. 10. Averages ± SEs of maximal aerobic power as a function of average blood hemoglobin concentrations in normal and marginally malnourished Colombian boys from study I (reprinted with permission from Spurr [24]).

(p < 0.001) in the normal boys with the U children having values which closely follow the 50th percentile of the US data, while the urban control L boys follow approximately the 15th percentile, except in the oldest age group and the rural control group which are closer to the 5th percentile. The differences between the normal urban and rural L boys were statistically significant (F = 6.30; p = 0.013) with the rural boys having lower values than urban children. These urban-rural differences persisted in the 3-way ANOVA of all the lower socioeconomic children. Furthermore, there are nutritional group effects, with the nutritionally deprived children having lower blood hemoglobin values than the normal boys (fig. 9). There were no statistically significant differences between L Wt-Age and L Wt-Ht groups in the ANOVA [2].

Figure 10 shows the lack of relationship of $\dot{V}_{O_2}$ max (per kg BW) to blood hemoglobin concentration in the range of 9.4–15.7 g/dl. The correlation coefficient between the two values (n = 1,008) was r = 0.03 (p = 0.14). A correlation analysis was also done on the age-adjusted blood hemoglobin and aerobic power values [($\bar{x}$–x of normal boys for age group)/SD of $\bar{x}$) × 100]. The correlation coefficient was −0.04 and was not statistically significant [2].

Table 2. Hematologic measurements of subjects in study II

Age groups years	Boys				Girls				3-way ANOVA probability
	control		undernourished		control		undernourished		
	n	mean ± SD	n	mean ± SD	n	mean ± SD	n	mean ± SD	
Hemoglobin, $g \cdot dl^{-1}$									
6–8	35	12.6 ± 0.9	33	12.7 ± 1.1	23	12.8 ± 1.1	27	12.6 ± 1.1	A < 0.001
10–12	27	13.3 ± 0.8	53	13.2 ± 1.2	22	13.2 ± 1.0	30	13.2 ± 0.7	NG = n.s.
14–16	31	14.4 ± 1.4	37	13.5 ± 1.2	15	13.1 ± 1.1	19	13.5 ± 1.1	S = n.s.
									A × S = 0.04

ANOVA = Analysis of variance; A = age; NG = nutritional group; S = Sex; n.s. = not statistically significant.

Study II. The values for blood hemoglobin obtained in study II are presented in table 2. There were significant increases in hemoglobin with age ($p < 0.001$) but, contrary to the results obtained in study I (fig. 9), there was no statistically significant nutritional group effect, nor were differences in the sexes statistically significant (table 2). There was a statistically significant age-sex interaction ($p = 0.04$).

Aerobic Power of Children in Different Countries

Figure 11 compares the aerobic power of boys and girls in study II with values obtained in one developing [4] and two developed countries [15, 16]. The criteria for inclusion in the comparison were that the investigators made a careful effort to reach true maximum values, that the treadmill was the ergometer, that several age-group values of school-aged boys and girls were available and that maximum heart rates were also presented in the data.

It can be seen in figure 11 that, with one exception, the group averages of maximum heart rate were all over 200/min (201–211/min) and that there is really little difference in the values obtained in the three countries.

The maximum aerobic power of Dutch and Colombian children were almost identical and only slightly below those reported for Swedish children. Åstrand and Rodahl [17, pp. 401–402] quoted the results of Asmussen, who analyzed the $\dot{V}_{O_2}$ max data of Åstrand [15] in relation to height (H) of the subjects and found that it was proportional to $H^{2.9}$ in males 8–18 years of age and to $H^{2.5}$ in females 8–16 years of age. The same calculations for the data of the children in Study II were $H^{2.9}$ and $H^{2.9}$ for the control and undernourished boys and $H^{2.50}$ and $H^{2.57}$ for the control and undernourished girls, respectively.

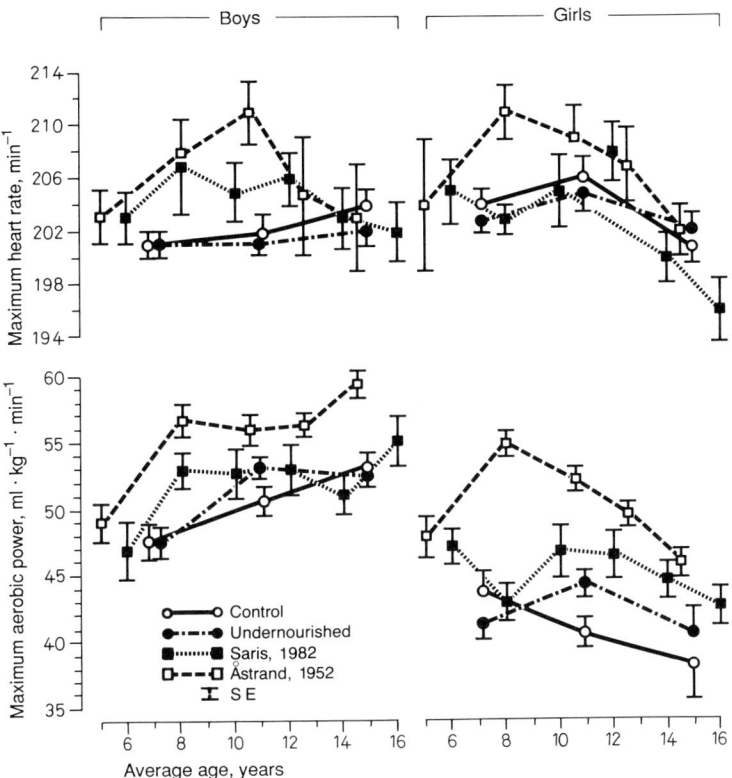

Fig. 11. Comparison of the $\dot{V}_{O_2}$ max (ml/kg · min body mass) data from study II of nutritionally normal and marginally undernourished Colombian children with those obtained on Dutch [16] and Swedish [15] boys and girls (reprinted with permission from Spurr and Reina [4]).

Discussion

Nutrition and Growth

During periods of active growth, two adaptive mechanisms available to an individual to balance a chronic marginal deficit in energy intake are decreased physical activity and/or decreased growth. Although the energy requirement for growth is not large, the shortness of adult stature in regions where chronic energy deficiency is endemic is well recognized and is due largely to environmental influences which include inadequate energy intake and chronic infections [18].

The process is described for our subjects in figures 1–4. The selection criteria employed in these studies to recruit subjects to fit predetermined categories was an attempt to separate nutritional from socioeconomic effects in the statistical analysis of the data. The children selected in this manner were indeed undernourished, as evidenced by the delay in the age of the growth spurt (fig. 5) and of sexual maturation (table 1), which are cardinal features of malnutrition [19, 20] and the fact that those classified as undernourished exhibited a slowed pattern of growth which will lead to stunting as adults. Furthermore, the undernourished also had reduced fat stores and lower values for LBM (fig. 2, 4).

The data for $\dot{V}_{O_2}$ max (liters/min; fig. 6) demonstrate clearly that both urban and rural children with a past history (L Wt-Age) or ongoing (L Wt-Ht) malnutrition have significantly lower values ($\sim 85\%$) than their nutritionally normal counterparts throughout the age range studied. A similar pattern of growth of $\dot{V}_{O_2}$ max with age was observed in normal and marginally undernourished boys and girls [4]. When the averages for the 35 groups of boys in study I and the 12 groups of boys and girls in study II are plotted on LBM the data fall on a straight line (figs. 7, 8) indicating that the observed differences are predominantly the result of differences in body size, as represented by the LBM. The line formed by the data obtained in the girls was lower and significantly different from that found in their male counterparts (fig. 8) indicating that they were in poorer physical condition. It has also been demonstrated in these same girls that their levels of activity as measured by their energy expenditure are lower than those in boys, perhaps as a result of cultural pressures on young girls to follow 'lady-like' patterns of behavior [21].

The data for aerobic power and maximum heart rates obtained in nutritionally normal and undernourished Colombian children were very comparable (fig. 11) with those measured in Dutch and Swedish boys and girls. Consequently, when the methodology is similar (treadmill), and care is taken in instructing young children in the procedure, $\dot{V}_{O_2}$ max testing can be carried out with reproducible results in subjects of varying ethnic and cultural backgrounds studied in different laboratories. Young children are in remarkably good physical condition with average values of aerobic power which exceed 50 ml/kg · min BW in most age groups (fig. 11). Furthermore, the procedure is safe when care is taken to examine subjects carefully. At this writing we have carried out this procedure in excess of 1,500 boys and girls without a single untoward incident.

In study I, it was found that urban boys had higher values for aerobic capacities per kg BW than rural boys [2]. This may be due to the ready availability of sports facilities and programs in the city of Cali which are not present in rural areas. Shephard et al. [22] encountered the same differences in urban-rural aerobic capacities in Canada and suggested that differences in access to recreational facilities could have contributed to the lower values found in rural children. This dispels the commonly held view that urban children are less physically active than rural children in modern Colombian society.

The lower total $\dot{V}_{O_2}$ max (liters/min) of the nutritionally deprived groups means that since the additional energy cost of doing tasks, over and above that required for body movement, is the same for both normal and L Wt-Ht boys [23], the latter have to expend a relatively greater effort (% $\dot{V}_{O_2}$ max) than the former in the performance of physically related tasks. It has been shown in adults that low $\dot{V}_{O_2}$ max is associated with low productivity in moderate to heavy work [24]. The implications of these findings of decreased work capacity and productivity in heavy physical work of these children when they reach adulthood have been discussed in a series of recent articles [3, 24–27].

The increase in blood hemoglobin concentration with age in children is a well-known fact [13] and was observed in all groups studied (fig. 9, table 2). Iron deficiency is the major cause of low hemoglobin levels in Latin America [28], and presumably played a role in the socioeconomic and nutritional group effects observed in study I (fig. 9). The reason for the lack of nutritional group differences in the later study II is not known. We have been unable to discover any interventions which might help explain the discrepancy between the two studies.

In adults, low blood hemoglobin concentrations are associated with reduced physical work capacity [29–31] and productivity [32]. Davies [33] also reported reduced work capacity in anemic 14-year-old boys (hemoglobin 8.0 ± 0.6 g/dl; mean $\pm$ SD). There appears to be a deleterious effect of even mild anemia on physical work capacity in adults [31]. There was no difference between anemic and nonanemic children in energy expenditure at play or in school (grades 1–3) either before or after iron therapy [34] but this may have been more related to the way the study was performed than to a lack of effect of the iron therapy itself. Vellar and Hermansen [35] have also made observations similar to those presented in fig. 10. Parsons and Wright [36] measured the heart rate response of anemic children to stair climbing and concluded that the adaptation to chronic anemia was so complete that practically no impairment of exercise tolerance resulted. These results indicate that in chronically anemic children there are possible adaptations in cardiac output, arteriovenous O_2 extraction, blood volume (total hemoglobin), 2,3-diphosphoglycerate levels, blood flow distribution, etc., which appear not to have been studied during exercise.

The mechanism of adaptation to low blood hemoglobin in children is an interesting physiological question. Its contribution, if any, to adult adaptability to low hemoglobin levels is unknown. The sensitivity of the work capacity of adult Guatemalan agricultural laborers to small decreases in blood hemoglobin [31] would suggest that there probably is no carryover. It is likely that these Guatemalan workers have passed through phases of low hemoglobin during childhood similar to those experienced by our Colombian subjects (fig. 9), since iron deficiency anemia is prevalent in Latin America [28] as it is in the rest of the developing world [37].

Conclusion

Studies of aerobic power in school-aged Colombian children 6–16 years of age, selected on the basis of national normative data as nutritionally normal or nutritionally deprived either at some time in the past, so that weight-for-age was below normal, or at the time of the studies based on low weight-for-age and low-weight-for-height, were carried out in over 1,000 subjects living in both urban and rural areas. The depression in the development of aerobic power associated with nutritional deprivation is related to the associated depression in growth (body size) which occurs as a consequence of chronic dietary energy deficiency. With the slowing of growth in these children there is a concomitant delayed growth spurt and sexual development.

The results of this slowed growth and smaller adult size will be a reduced physical work capacity which implies reduced productivity in heavy physical work. Individuals then, have been deprived of their maximum physical development potential and their society of their maximum economic contribution in those countries which may have a major dependence on heavy manual labor.

References

1 Spurr GB, Reina JC, Barac-Nieto M: Marginal malnutrition in school-aged Colombian boys: anthropometry and maturation. Am J Clin Nutr 1983;37:119–132.
2 Spurr GB, Reina JC, Dahners HW, Barac-Nieto M: Marginal malnutrition in school-aged Colombian boys: Functional consequences in maximum exercise. Am J Clin Nutr 1983;37:834–837.
3 Spurr GB: Marginal nutrition in childhood: implications for adult work capacity and productivity; in Collins JK, Roberts DF (eds): Capacity for Work in the Tropics. Cambridge, Cambridge University Press, 1988, pp 107–140.
4 Spurr GB, Reina JC: Maximum oxygen consumption in marginally malnourished Colombian boys and girls 6–16 years of age. Am J Hum Biol 1989;1:11–19.
5 Reuda-Williamson R, Luna-Jaspe H, Ariza J, Pardo F, Mora JO: Estudio seccional de crecimiento, desarrollo y nutrición en 12, 138 niños de Bogotá, Colombia. Pediatría 1969; 10:337–349.

6 Pařízková J: Total body fat and skinfolds in children. Metabolism 1961;10:794–807.
7 Barac-Nieto M, Spurr GB, Reina JC: Marginal malnutrition in school-aged Colombian boys: body composition and maximal O_2 consumption. Am J Clin Nutr 1984;39: 830–839.
8 Brožek J, Grande F, Anderson JT, Keys A: Densitometric analysis of body composition: revision of some quantitative assumptions. Ann NY Acad Sci 1963;110:113–140.
9 Tanner JM: Growth as a monitor of nutritional status. Proc Nutr Soc 1976;35:325–332.
10 Balke B, Ware W: An experimental study of physical fitness of Air Force personnel. US Armed Forces Med J 1959;10:675–688.
11 Hamill PVV, Drizd TA, Johnson CL, Reed RB, Roche AF, Moore WM: Physical growth: National Center for Health Statistics percentiles. Am J Clin Nutr 1979;332: 607–629.
12 Spurr GB, Reina JC, Barac-Nieto M, Maksud MG: Maximum oxygen consumption of nutritionally normal white, mestizo and black Colombian boys 6–16 years of age. Hum Biol 1982;54:553–574.
13 Garn SM, Ryan AS, Abraham S, Owen G: Suggested sex and age appropriate values for 'low' and 'deficient' hemoglobin levels. Am J Clin Nutr 1981;34:1648–1651.
14 Garn SM, Ryan AS, Owen GM, Abraham S: Income matched black-white hemoglobin differences after correction for low transferrin saturations. Am J Clin Nutr 1981;34: 1645–1647.
15 Åstrand P-O: Experimental Studies of Physical Working Capacity in Relation to Sex and Age. Copenhagen, Munksgaard, 1952.
16 Saris, WHM: Aerobic Power and Daily Physical Activity in Children. Meppel, Kripps Repro, 1982.
17 Åstrand P-O, Rodahl K: Textbook of Work Physiology. New York, McGraw-Hill, 1986.
18 Martorell R: Child growth retardation: A discussion of its causes and its relationship to health; in Blaxter K, Waterlow JC (eds): Nutritional Adaptation in Man. London, Libbey, 1985, pp 13–29.
19 Viteri FE, Torún B: Protein-calorie malnutrition; in Goodheart RS, Shils ME (eds): Modern Nutrition in Health and Disease. Philadelphia, Lea & Febiger, 1980, pp 697–720.
20 Tanner JM, Whitehouse RH: Clinical longitudinal standards for height, weight, height velocity, weight velocity and stages of puberty. Arch Dis Child 1976;51:170–179.
21 Spurr GB, Reina JC: Patterns of daily energy expenditure in normal and marginally undernourished school-aged Colombian children. Eur J Clin Nutr 1988;42:819–834.
22 Shephard RJ, Lavallée H, Larivière G, Rajic M, Brisson GR, Beaucage C, Jéquier J-C, LáBarre R: La capacité physique des enfants canadiens: Une comparaison entre les enfants canadiens-français, canadiens-anglais et esquimaux. I. Consommation maximale d'oxygène et débit cardiaque. Union Méd Can 1974;104:1767–1777.
23 Spurr GB, Reina JC: Marginal malnutrition in school-aged Colombian boys: body size and energy costs of walking and light loadcarrying. Hum Nutr Clin Nutr 1986;40C: 409–419.
24 Spurr GB: Nutritional status and physical work capacity. Yearb Phys Anthropol 1983; 26:1–35.
25 Spurr GB: Physical activity, nutritional status and physical work capacity in relation to agricultural productivity; in Pollitt E, Amante P (eds): Energy Intake and Activity. New York, Liss, 1984, pp 207–261.
26 Spurr GB: Effects of chronic energy deficiency on stature, work capacity and productivity; in Schürch B, Scrimshaw NS (eds): Chronic Energy Deficiency: Consequences and Related Issues. Lausanne, Nestlé Foundation, 1988, pp 95–134.

27 Spurr GB: Body size, physical work capacity, and productivity in hard work: Is bigger better? in Waterlow JC (ed): Linear Growth Retardation in Less Developed Countries. New York, Raven Press, 1988, pp 215–243.
28 Cook JD, Alvarado J, Guthisky A, Jamra M, Labardini L, Layrisse M, Linares J, Loria A, Maspes V, Restrepo A, Reynafarje C, Sanchez-Medal L, Velez H, Viteri F: Nutritional deficiency and anemia in Latin America: A collaborative study. Blood 1971; 38:591–603.
29 Barac-Nieto M, Spurr GB, Maksud MG, Lotero H: Aerobic work capacity in chronically undernourished adult males. J Appl Physiol 1978;44:209–215.
30 Davies CTM, Chukweimeka AC, van Haaren JPM: Iron deficiency anemia: Its effect on maximum aerobic power and responses to exercise in African males aged 17–40 years. Clin Sci 1973;44:555–562.
31 Viteri FE, Torún B: Anaemia and physical work capacity. Clin Haematol 1974;3:609–626.
32 Basta SS, Soekirman, Karyadi E, Scrimshaw NS: Iron deficiency anemia and productivity of adult males in Indonesia. Am J Clin Nutr 1979;32:916–925.
33 Davies CTM: Physiological responses to exercise in East African children. I. The effects of shistosomiasis, anemia and malnutrition. J Trop Pediatr Environ Child Health 1973; 19:115–119.
34 Gandra YR, Bradfield RB: Energy expenditure and oxygen handling efficiency of anemic school children. Am J Clin Nutr 1971;24:1451–1456.
35 Vellar OD, Hermansen L: Physiologic performance and hematologic parameters. Acta Med Scand 1971;(suppl 522):1–30.
36 Parsons CG, Wright FH: Circulatory function in the anemias of children. I. Effect of anemia on exercise tolerance and vital capacity. Am J Dis Child 1939;57:15–28.
37 Scrimshaw, NS: Functional consequences of iron deficiency in human populations. J Nutr Sci Vitaminol 1984;30:47–63.

G.B. Spurr, PhD, Department of Physiology, Medical College of Wisconsin,
8701 Watertown Plank Road, Milwaukee, WI 53226 (USA)

Growth, Nutrition and Physical Performance in Algeria[1]

N. Dekkar

National Institute of Public Health, Algiers, Algeria

Introduction

Numerous developing countries have experienced profound transformations of health status over the last three decades, with substantial decreases of infant mortality and a continued increase of life expectancy among their populations. But we still lack large, representative samples describing the nutritional status and physical development of such populations. The Algerian situation provides a good illustration of this point.

A knowledge of the environment in which these children live is helpful to an understanding of the data. The Algerian population is currently about 50% rural and 50% urban, but the rate of exodus from the rural setting is rapid. The population density diminishes rapidly on moving towards the south of the country. Among principal demographic indices, we may note that the infant mortality decreased from 130 per 1,000 to less than 50 per 1,000 between 1970 and 1989, while the annual rate of population growth decreased from 3.21 to 2.70%, and life expectancy at birth increased from 53.4 to 65.0 years over the same period. Infectious diseases are the primary cause of death. Those under 20 years of age comprise 58% of the total population. The family size is large, averaging 7 living children. In our sample, 53% of the fathers could read, a figure above the national average.

The number of physicians increased from 1 per 5,500 inhabitants in 1967 to 1 per 1,200 inhabitants in 1989, but doctors are still unequally distributed across the country.

The largest changes have been in the economic sector. The domestic consumption increased from 4 billion US dollars in 1967 to 20.3 billion US dollars in 1983. The gross domestic products (GDP) for those same years

---

[1] Translated and abridged by R.J. Shephard.

were, respectively, US $265 and US $2,264 per inhabitant. Over this same interval, the ratio of domestic consumption to GDP increased to 50.3%. The national health expenditures rose from 2% of the GDP in 1973 to 5.5% in 1988. Since 1985, health expenditures have grown more rapidly than the GDP. In its annual report of 1986, UNICEF classed Algeria as a moderately developed country, with a gross national product of US $2,590 per inhabitant.

Thus, all the conditions described by Morley [1] for the development of pediatric problems in developing countries exist in Algeria. All of the statistics studied, with the exception of current patterns of education, underline the existence of an unfavorable environment for optimization of health. The students studied here were born between 1963 and 1978, growing up in large families with (on average) a low level of education, a low income and crowded housing. The high infant mortality rate has exercised a selective pressure on this sample. The surviving children have grown up in an unfavorable social, nutritional and medical environment. However, a comparison of representative samples from cohorts born between 1955 and 1959 and 1969 and 1973 demonstrates a frank progression of both nutritional and anthropometric parameters.

The information on diet and global nutrition lacks precision. The diet is essentially based on carbohydrates (66% of energy needs), with a predominance of grains (wheat and barley). Both the total daily energy intake per inhabitant and the nutritional characteristics of the food have changed between 1967 and 1987 [2]. In 1987, the average energy intake of energy was 12.6 MJ (3,000 kcal/day), whereas it had previously remained very stable at 11.3 MJ, 2,700 kcal/day for 13 years.

The proportion of fat in the diet increased progressively from 1967 to 1987, meeting from 14 to 22% of energy needs as compared with the recommended value of 25%. The proportion of proteins remained at the relatively low level of 12%, while the proportion of carbohydrates, although decreasing, remained above the recommended level. Such proportions of protein, fat and carbohydrate seem characteristic of each of the three normal meals. However, there are substantial variations with region of residence, socioeconomic status and season of the year. It is estimated that 30% of the population consumes 60% of the available animal protein. In 1973, 26.3% of children between birth and 36 months who were old enough to eat fruit and vegetables had never had the opportunity to do so; in 1981, this was still true of 14.2% of such children. Looking at the same age group, 40.3% of children in 1973 and 22.1% in 1981 had never had the opportunity to eat animal protein.

The available studies [3], although showing certain disparities and methodological weaknesses, all tend to indicate a qualitative insufficiency of nutrition for both the nursing mother and the young child, with available food unevenly distributed across the country and between social classes. At the same time, the data show an improvement of nutrition between the first and the last cohorts of our study.

Methods and Materials

Methods

After several preliminary trials, we carried out a major cross-sectional study of students in 1983. Random cluster sampling was based on communities, schools and individual school classes, stratified according to age, sex, socioeconomic status and rural or urban residence. This type of survey has the advantage that it can easily be reproduced [4]. In all, data were collected from 38 schools spread across 10 communities.

The variables studied included socioeconomic status, geoeconomic stratum, health status, anthropometry and physical performance. Where possible, standard methods were used, for example: (i) the classification of the British Registrar General for social class [5]; (ii) international conventions for anthropometry [5]; (iii) the methods and techniques of WHO for exercise testing; (iv) the tests of AAHPER-D or the IBP [8, 9] for physical performance, and (v) the techniques of FAO, WHO and UNO for energy expenditure.

Information was collected by ten teams of observers. Each was directed by a doctor, and comprised also a health technician, two exercise specialists and two teachers. The age of the students was expressed in completed years. Among measures taken to ensure the validity of data, we may note the choice of universally accepted tests, the meeting of Waterlow's criteria [10], the technical expertise and knowledge of the observers, a standardization of techniques between the ten teams of observers, an examination of reproducibility (comparing differences between observers), and careful control of data entry (0.26% of errors in the final sampling). Intergroup comparisons were generally broken down by age and sex. Because of multiple comparisons, note was taken of the proportion of 'significant' differences, and the consistency in the direction of change.

Materials

The sample comprised 5,603 boys and 5,358 girls aged 5–20 years, with only a limited representation of students at the extremes of the age distribution. Some 51% of students were drawn from urban and 49% from rural areas. Because of the limited numbers of students in the upper social categories, social classes 1 and 2 were combined to yield class A (5.1% of the sample), classes 3 and 4 formed class B (62.9%), and class 5 became class C (32%).

Three children out of 10 took at least one meal per day at school, this being particularly true of children aged 5–11 years. About a quarter of the boys and a third of the girls were not involved in any type of physical activity or sport. Only 2 of 3 students participated in physical education or sport at school (effectively, 1 h/week), and less than 5% played sport outside of the school. The practice of sport was more common among boys than girls, and it increased with age.

Clinical examination disclosed 4.8% with medical anomalies. Most were cardiovascular and respiratory disorders without great functional significance, but handicaps compatible with school attendance were observed in 2.2% of the subjects. Children with anomalies were excluded from the anthropometric measurements and the tests of physical performance in order to avoid biassing the results.

Results

Standing Height

In boys, the average stature increased regularly (r = 0.99) from 109.9 cm at 5 years to 170.6 cm at 18 years. An acceleration of growth was observed between 12 and 14 years, with a progressive slowing to plateau values thereafter. The standard deviation increased progressively to 14 years, but declined thereafter (fig. 1). In girls, height again showed a linear correlation with age (r = 0.96). Mean values increased from 109.5 cm at 5 years to a maximum of 157.5 cm at 18 years. Growth spurts (6–7 years, 9–10 years, 11–12 years) alternated with periods of reduced growth (5–6 years, 7–9 years, 10–11 years), the largest acceleration of growth occurring between 9 and 10 years. The standard deviations showed an inverted parabola, with maximum variability at 12 years (fig. 1).

Comparing data by zone of residence (urban vs. rural), the boys showed no significant differences in either mean values or variability except at 15 years (p<0.01). Likewise, in the girls the only significant differences were for the mean heights at 10 and 12 years. Height was not statistically related to geoeconomic stratum in 68.8% of comparisons for the boys; significant differences (8, 9, 11, 14 and 19 years) did not show any particular age concentration. In the girls, geoeconomic stratum was a significant influence in 50% of the comparisons, with a concentration of such differences between 6 and 11 years. Neither sex showed a constant predominance of any one stratum. Growth curves separated by geoeconomic stratum were thus similar to those previously described. A comparison of paired variances for socioeconomic status (A/B, B/C, A/C) revealed significant differences at most ages, with the exception of comparisons between classes B and C in the girls. In the boys, the difference of means between the three social classes increased progressively with age. However, in the girls, there were few significant differences, except that between 11 and 12 years class A showed an earlier acceleration of growth.

A comparison of mean values from the present study with those obtained on boys of 11–14 years by Sprynar and Sprynarova [11] in 1969/1970 shows an average advantage of 5.5 cm to the current sample. A comparison with Chamla and Demoulin's [12] results for both sexes leads to a similar conclusion. There is no significant difference from the mean values observed in Tunisia [13]. On the other hand, a comparison with the NCHS norms [10] shows that Algerian students of both sexes are smaller from the age of 5 years, and the difference widens with age, because puberty and the growth spurt occur later. The variability of data is also greater in Algeria. In general, the Algerian means are about 1 SD below the international norms, and are never more than 1.5 SDs below such norms (fig. 1).

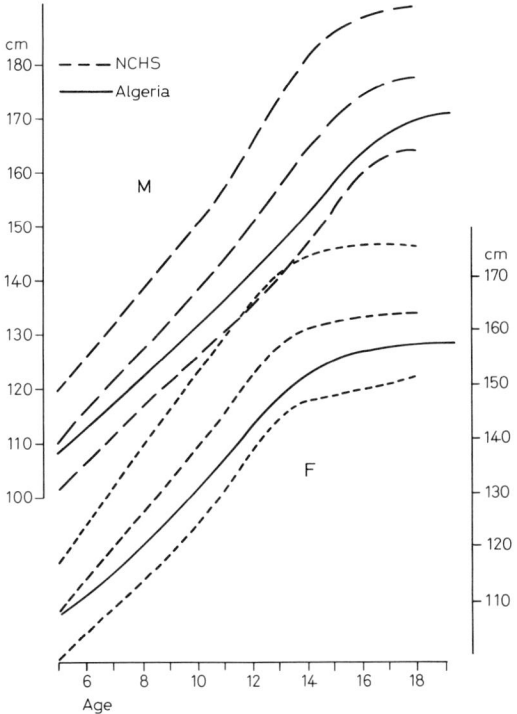

Fig. 1. Relation of stature to age (years). A comparison of mean data for Algerian boys (M) and girls (F) with norms proposed by the US National Center for Health Statistics.

Body Mass

The relationship of body mass to age has limited interest since Keller and Fillmore [14] have shown that this relationship can be explained by the relationships of mass to height and of height to age. Nevertheless, several authors from different countries have expressed their data in this manner.

In the boys, there is a linear growth of body mass from 6 to 13 years, followed by an acceleration to 18 years, after which the data tend to plateau. The standard deviations increase progressively to 15 years, and diminish thereafter (fig. 2). In the girls, the growth curve shows four phases – a constant progression from 6 to 9 years, an acceleration to 12 years, a second increase of slope at 13 years, and a relative stabilization at 16 years. The variability increases rapidly to reach a maximum at 13 years, and stabilizes thereafter (fig. 2).

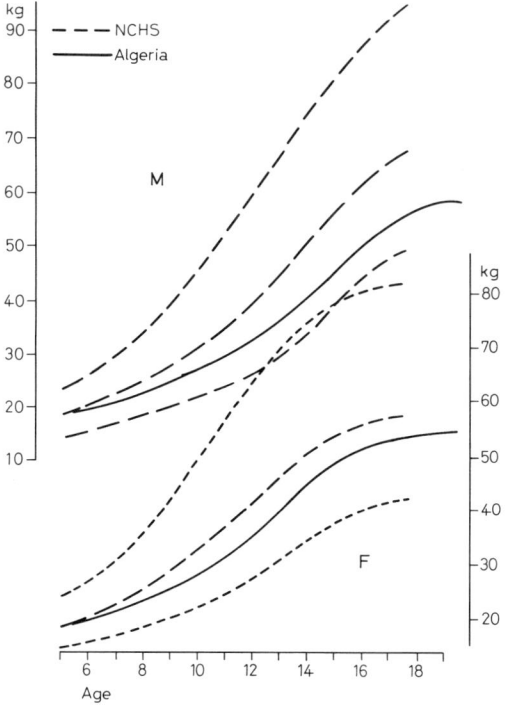

Fig. 2. Relation of body mass to age (years). A comparison of mean data for Algerian boys (M) and girls (F) with norms proposed by the US National Center for Health Statistics.

The socioeconomic status apparently has no influence on the variability of the data, but does affect the means; the higher the socioeconomic status of the boys, the greater their mass. Comparing body mass by zone (urban vs. rural), the boys show an equality of variances in 60% of age comparisons, but a difference of means in 53% of comparisons. In the girls, both means and variances are equal in two-thirds of comparisons. Thus, the mean masses do not differ between urban and rural communities, except in boys aged 6–15 years. The total body mass is the sum of lean tissue and fat mass, and the latter is significantly increased in the urban students of both sexes.

In the majority of age comparisons (9 of 14), the boys show no difference of body mass when classified by geoeconomic stratum. In the girls, also, 86% of age comparisons are unaffected by geoeconomic stratum, but there are differences in the prepubertal and pubertal periods; no stratum is uniquely different, with the possible exception that in the majority of age groups the mass are lowest in those students from the least-favored stratum.

Comparing the 1983 results with data for 1970 [11] and 1974 [12], mean masses have increased. The largest gain (3.5 kg) has been at 14 years, and there has also been an increase in the variability of the data. The Tunisian population (which is closely similar from a geo-ethnic point of view) shows very similar mean values to current Algerian data [13]. Comparing results with international norms [10], mean values in the boys pass from -0.5 to -1.5 SDs as age increases (fig. 2). In the girls, values remain between the median and -0.5 SD at all ages.

Mass/Height Ratios

The ratio of mass to height has been determined (in 1-cm steps) between 100 and 180 cm for boys, and between 100 and 170 cm for girls (fig. 3). The correlation between body mass and stature for each pair of values is very high for the total sample ($r = 0.92$). For a given age and sex, r remains highly significant, at about 0.67.

A comparison of the mass/height ratios with the 1970 values for boys aged 11–14 years [11] shows no difference between the two samples. Comparing values to the NCHS medians, there is again a remarkable identity of values from 100 to 145 cm in the boys, and from 100 to 137 cm in the girls (fig. 3). The entire span of values can also be compared with French norms [6], values being comparable for the boys, and slightly above the French norms for the girls, with an increase of variability.

Dietary Needs

The minimal needs of energy and protein are summarized in table 1. The basal metabolism per kilo of body mass is higher than the international norms, because of the shorter stature of the Algerian student. On the other hand, the daily energy needs and protein requirements at any given age are somewhat reduced by the low body mass.

Vital Capacity

The vital capacity increases progressively with age, in boys from 1.5 liters at 10 years to 3.5 liters at 20 years (fig. 4), and in girls from 1.4 to 2.5 liters (fig. 4). These values show a substantial correlation with body mass. By the age of 20 years, the vital capacity is 56.9 ml/kg in males, and 47.6 ml/kg in females.

A comparison of vital capacities by zone of residence (urban vs. rural) does not show many differences between 11 and 20 years; even where there are apparently significant differences, they are inconsistent in direction. The mean vital capacity at almost all ages is again apparently influenced by the

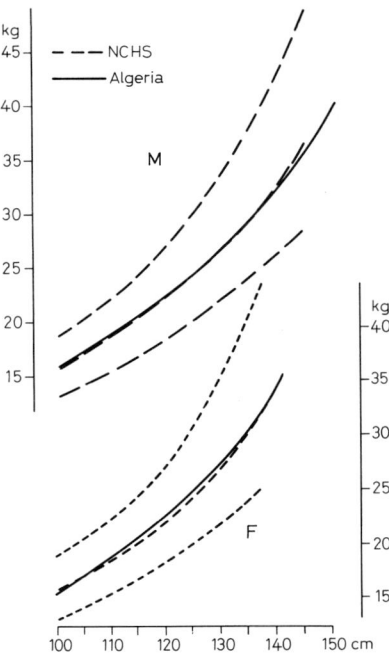

Fig. 3. Body mass as a function of standing height. A comparison of mean data for Algerian boys (M) and girls (F) with norms proposed by the US National Center for Health Statistics.

Table 1. Daily average energy requirements and safe level of protein intake for adolescents age 10–18 years

Sex	Age years	BMR/kg		Daily energy requirement			Safe level of protein intake	
		kcal	kJ	BMR factor	kcal	kJ	g/kg	g/day
M	10–12	38.9	162.4	1.76	2,100	8,800	1.0	30
	12–14	35.4	147.9	1.69	2,200	9,200	1.0	36
	14–16	31.6	132.1	1.65	2,400	10,500	0.95	43
	16–18	29.3	122.4	1.62	2,600	10,900	0.90	48
F	10–12	35.9	149.9	1.63	1,850	7,750	1.0	32
	12–14	30.8	128.6	1.58	2,000	8,400	0.95	39
	14–16	27.6	115.2	1.55	2,100	8,800	0.90	44
	16–18	26.4	110.3	1.52	2,100	8,800	0.80	42

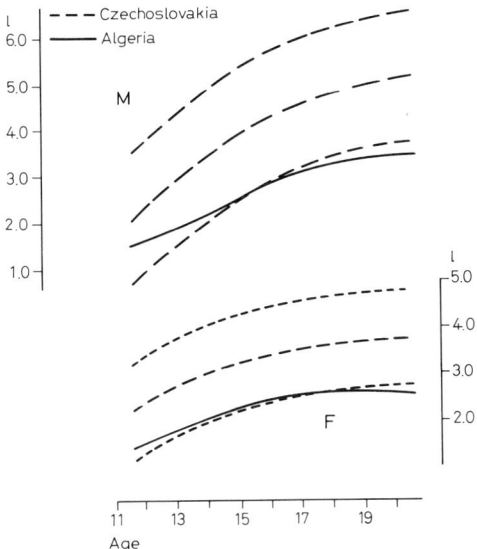

Fig. 4. Relationship of vital capacity (liters BTPS) to age (years). A comparison of mean data for Algerian boys (M) and girls (F) with norms for the Czechoslovak population obtained by Seliger.

socioeconomic status, but no one socioeconomic stratum is uniformly superior or inferior at all ages. Boys from the highest socioeconomic status seem to have the highest vital capacities, but because of the small number of students falling into this age category, the effect is only statistically significant in 40% of age comparisons. Even if the same proportion of significant differences had been observed in the girls, it would be difficult to speak of an advantage to one social class.

A classification of vital capacity by practice of organised sport shows that the active students have higher values in most age categories. On the other hand, there is no difference between boys who participate in a school physical education programme and those who do not. In the girls, also, the vital capacity is increased by the practice of sport at most ages.

The vital capacity of the Algerian students is substantially inferior to that of Czech students, whether results be expressed as absolute volumes or per kg of body mass (fig. 4).

Physical Work Capacity

In the boys, the PWC_{170} is linearly related to age ($r = 0.93$; fig. 5). In girls, there is a weaker linear relationship (fig. 5). Both sexes show a

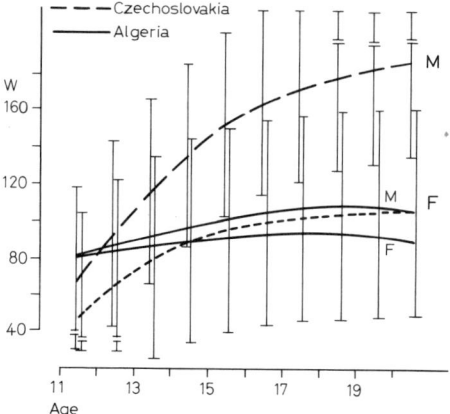

Fig. 5. Absolute PWC$_{170}$ in relation to age (years). A comparison of mean data for Algerian boys (M) and girls (F) with values for the Czechoslovak population reported by Seliger.

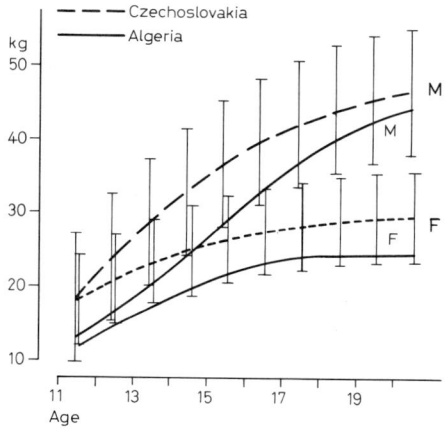

Fig. 6. Right-hand grip (expressed in kg; the value in Newtons would be 9.81 times larger) in relation to age (years). A comparison of mean data for Algerian boys (M) and girls (F) with values for the Czechoslovak population reported by Seliger.

substantial variability (11–23%). If data are expressed per kg of body mass, the relationship to age becomes negative ($r = -0.91$ for girls, -0.96 for boys), with values decreasing from 2.46 ± 0.47 W/kg at 11 years to 1.70 ± 0.27 W/kg at 20 years in both sexes.

Urban/rural differences in the boys were not significant. In the girls, there were only differences between 14 and 16 years, these favoring the rural girls. The boys showed apparent differences with geoeconomic stratum at most ages, but without obvious predominance of one stratum. In the girls, differences were only significant at 14 and 15 years. Likewise, there were no effects of socioeconomic status in the boys, and none in the majority of ages for the girls. Participation in organised physical activity was also without apparent influence on the PWC_{170}.

Comparison with results from other countries [16] showed an advantage of physical working capacity to the Algerian boys before the age of puberty. However, after puberty, the overseas values were larger. The Algerian girls had a larger PWC than their peers in Romania, francophone Canada and Belgium [15] until the age of 14 years, but in the older age categories values were comparable with these other samples. In contrast, Czech students had a larger PWC_{170} at all ages except 11 years [15]; in the Czech sample, there was also a positive relationship between age and W/kg.

Handgrip Force

The right handgrip force was linearly related to age in both sexes (r = 0.98). The growth curve showed an inflexion around 13 years for the boys (fig. 6), and 12 years for the girls. The coefficient of variation decreased from 50% at 7 years to about 25% at 18 years. The curve for the left hand followed a generally similar pattern, although mean values were slightly lower in both sexes.

Urban/rural comparisons showed significant differences of mean values only in boys of 16 and 19 years. In the girls, there were significant differences in about a half of the age groups, particularly between 14 and 18 years, students from the rural areas showing the larger grip forces. A significant influence of geoeconomic stratum was demonstrated at all ages, the richest region having the lowest grip scores, and the poorest region the highest scores; however, for the intermediate strata, differences changed direction with age. In contrast, most age groups did not show an effect of socioeconomic status; in the 30% of comparisons where differences were significant, the direction of change was inconsistent.

In the boys, 60% of the comparisons by practice of sport were significantly different, but the direction of change was inconsistent, so that it was not possible to claim that the sport had led to a larger grip strength. Moreover, active and inactive groups had similar scores between 15 and 18 years of age. In the girls, only a single comparison was significant at 12 years.

In both sexes, the Algerian scores were poorer than those reported by Seliger and Bartunek [15] for Czech students, although the differences between the two samples tended to decrease with age (fig. 6).

Standing Long Jump

Beginning with a value of 129.7 cm at 11 years, the mean score for the boys increased in a quasi-linear fashion (r = 0.99) to a value of 175 cm at an age of 18 years (fig. 7). In the girls, the data also showed a linear growth (r = 0.90), from a mean of 119.7 cm at 11 years to 133.3 cm at 18 years, with a large gain in performance at 14 years. The variability of data ranged from 12 to 18%.

Urban/rural comparisons showed a larger effect in boys than in girls. In the boys, 70% of age comparisons were significant, favoring the urban environment. In girls, 20% of comparisons differed significantly, again favoring the urban environment. The geoeconomic stratum showed significant differences in both sexes at all ages; the richest stratum was uniformly superior to the other four strata, although there were no consistent differences within the remaining categories. The effect of socioeconomic status was very marked among the boys at all ages, with the wealthiest students having the highest scores. However, there was only an effect in 20% of comparisons in the girls, at 19 and 20 years.

The practice of sport led to significant differences in 70% of age categories, without emergence of a clear advantage to the active category.

Comparison with the Algerian data of 1970 [11] showed that the impact of socioeconomic status was currently more important than in the earlier sample (p < 0.001), with a generally analogous variability at all ages. Comparison with the data of Seliger and Bartunek [15] showed important differences in both sexes at all ages, the Algerian data tending to be some 2 SDs smaller than the Czech scores.

Sit-Ups

In the boys, data were linearly related to age (r = 0.94), with a phase of growth to 16 years, followed by a stabilization of scores (fig. 8). Coefficients of variation ranged from 23.2 to 38.4%. In the girls, the linear relationship (r = 0.91) showed only a shallow slope (fig. 8), with a very large coefficient of variation (43.0–69.9%).

Urban and rural growth curves were similar, although in the boys most age comparisons favored those from the urban environment. In the girls, most age categories showed no urban/rural differences. Boys with a high socioeconomic status had better scores through 13 years of age, but their advantage was not maintained in older age categories. In girls, differences were not significant from 11 to 14 years, but in the higher ages, those from the higher socioeconomic categories showed the higher scores.

In the boys, classification of data by the practice of sport left a linear relation of score to age (r = 0.68–0.98). In the girls, a relationship of score to age was only apparent in those who practiced physical education at

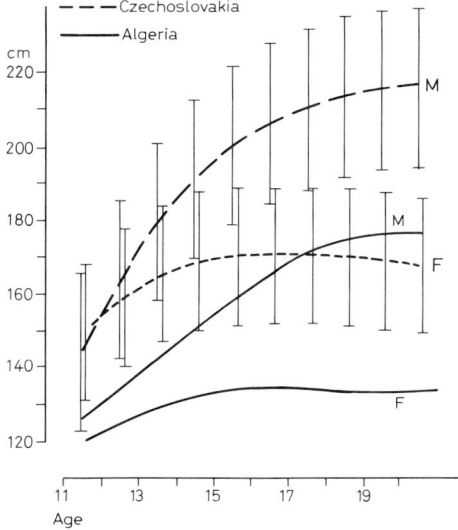

Fig. 7. Standing long jump in relation to age (years). Mean values for Algerian boys (M) and girls (F) compared with values reported by Seliger for the Czechoslovak population.

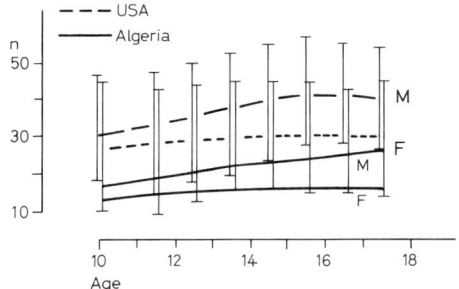

Fig. 8. Sit-ups (total number) in relation to age (years). Mean values for Algerian boys (M) and girls (F) compared with values reported for the US population (AAHPER, 1976).

school. In the boys, there was a significant advantage to the active students, but only from the age of 14 years. In the girls, most age comparisons favored those involved in sport.

Comparison of data with the results obtained in 1970 [11] showed that current means were actually lower, by about 10 sit-ups/min; this discrepancy cannot be entirely explained by an altered arm position (behind the neck in 1970, and on the opposite clavicle in 1983). Current results are similarly inferior to the norms for North America (fig. 8).

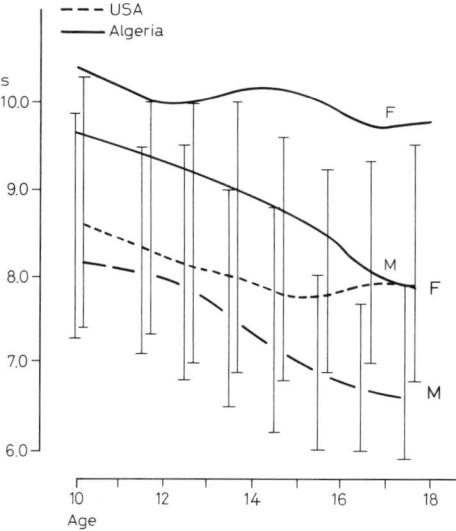

Fig. 9. Time required for the 50-meter dash. Mean results for Algerian boys (M) and girls (F) compared with norms reported for the US population (AAHPER, 1976).

Fifty-Meter Dash

In the boys, times for a 50-meter dash decreased from 9.4 s at 11 years to 7.3 s at 20 years (fig. 9). In the girls, corresponding times were 10.0 and 9.5 s (fig. 9). Both sexes reached a maximum velocity at the age of 19 years. Coefficients of variation ranged from 11 to 19%.

Urban/rural comparisons in the boys showed significant differences in about 50% of the age groups, favoring the urban environment. In the girls, differences did not differ in 70% of the comparisons, but in the remaining 30% differences again favored the urban milieu. The geoeconomic status significantly affected scores in all age categories for the boys and in 90% of age categories for the girls, although it was not possible to identify a consistently superior stratum. Likewise, in the case of socioeconomic status, the only significant difference in the boys was at 11 years, and there were no significant differences in the girls.

Significant differences related to sport involvement were found in 70% of age categories, the boys practicing sport having faster times than the other two categories of student. On the other hand, those participating in physical education at school gained no advantage over those who did not. In the girls, 50% of differences were significant, but there was no consistent relationship to either sport participation or participation in physical education at school.

A comparison with the 1970 data [11] showed a significant difference only at 13 years. The speed of the Algerians of both sexes was low relative to US norms (fig. 9).

Discussion

Nutritional Status

The current data on national food intake represents only the mean values for subjects of both sexes and all ages, but it does suggest that daily energy needs are satisfied, that minimal protein requirements are met, and that the proportions of nutrients (although still less than optimal) are moving in the right direction. In particular, lipids are progressively replacing carbohydrates. At any given age, the Algerian student is smaller than the international norms, and in consequence is also lighter. The mass/height ratio nevertheless corresponds with that of other populations, showing the remarkable stability of this index provided that the nutritional status of a community is adequate.

The data correspond to other reports in showing that the variables studied show different and irregularly placed rhythms of growth, sometimes with accelerations in the immediate prepubertal period, and sometimes in the pubertal period itself. In accordance with expectation, about 2.3% of the population fell below the critical value of (mean−2 SD) in terms of mass/height ratio, but in terms of height for age, 10–28% were more than 2 SDs below published norms at various ages. The normal mass/height ratio, but a low height for age and low mass for age suggest that the students previously suffered from protein/energy deficiency, even though they are currently well nourished [10].

Environmental Influences

Our study examined both macrosocial (urban/rural differences) and microsocial indicators (socioeconomic status, practice of sport and physical education).

In the boys, certain variables were larger among the urban samples. However, the urban/rural distinction must be interpreted with caution in recently independent countries, because of the rapid exodus from rural areas and the implantation of social and economic infrastructures in isolated parts of the country. Moreover, families remaining in rural areas have now largely abandoned agricultural pursuits, and are involved mainly in secondary and tertiary economic activities.

The most favorable geoeconomic strata do not show the largest scores, with the exception of scores for the explosive force of the lower limbs.

Individual regions may be more homogenous from a genetic and social perspective, but economic development of the favored regions has been too recent to have a detectable influence on health and hygiene. The influence of the geoeconomic stratum thus seems relatively uniform across the country, although the data cannot rule out local pockets of malnutrition.

The socioeconomic status is linked more closely to the nutritional status, with the highest classes showing the best scores. Algerian society, in common with other developing countries, has undergone profound change since independence, but the effects of such a change have yet to stabilize; even the most favored students have not yet reached the scores seen in developed countries.

The impact of physical education and sport is limited, but this seems a criticism of current practice (what effect could be expected from only 1 h per week?) rather than a lack of potential benefit from such pursuits.

Secular Changes

Despite a lessening of selective pressures as infant mortality has decreased, the average height and body mass of Algerian students has increased since the 1970s. Improved nutrition seems responsible for these gains, as witnessed by the smaller number of cases of malnutrition reported or found and the increased family expenditures on food. Better nutrition first augments the mass/height ratio, but later increases height for age (a gain of 4 cm over a period of 10 years). The residual discrepancy from international norms is age-related, with younger students (who have lived under improved circumstances) being more closely matched to accepted norms. The current gains are nevertheless precarious, and could be reversed by a period of drought or economic hardship.

Impact on Health

Under the influence of these environmental changes, the typical silhouette of the Algerian student has become longer and heavier, becoming like the Tunisian, although remaining lighter than the European or the North American. The age of puberty has also advanced since 1974 [12], matching that of Tunisians [13], although remaining behind that of Europeans [6]. The variability of data is still larger than in developed countries, emphasizing residual inequalities of the Algerian environment.

The ultimate goals of development are to improve the health and comfort of humankind, assuring the realization of genetic potential. Mortality and morbidity continue to fall, and children continue to become larger. But is such growth a desirable goal? To be taller or heavier is of no interest unless it is associated with greater physical, social or mental well-being. Do the observed anthropometric changes lead to an improvement of physical performance, and particularly those components of performance linked to good health?

The explosive strength of the lower limbs has increased, but the speed of movement has not improved, and the endurance of the trunk flexor muscles has apparently decreased. We thus suspect the impact of an increased burden of body fat and a lack of adequate physical activity. Improved nutrition should be linked to an increased energy expenditure and thus gains of cardiorespiratory performance rather than an accumulation of body fat. Although the improvements of nutrition to date have had incontestable and profoundly beneficial effects on the health of young people, certain excesses of nutrition may now be favoring the development of new pathologies. Can such risk factors be reduced by moderating food consumption and developing adequate and sustained programmes of physical activity?

Policy Implications

Developing countries like Algeria share many common problems such as a high infant mortality rate, a rapid demographic growth, low levels of education and limited family incomes. The current retardation of growth seems likely to diminish through the combined influences of improved nutrition, education and hygiene. Nevertheless, it is important to stress that nutrition is an 'output' of the social system, and is not closely tied to the available economic and sanitary resources. In Algeria, the improvement of nutritional status has not been uniform; currently, malnutrition co-exists with excess body mass. The optimal diet should be chosen to meet the requirements of growth and development, without reaching levels that have a negative impact on health or performance.

As countries develop, demographic and sanitary indicators become insufficient to judge health status. Nutritional studies become necessary. Cross-sectional analyses should look not only at height for age and mass for age, but also skinfold thicknesses and physical performance measures linked to health [9, 10]. It is also desirable to evaluate periodically the frequency, duration and intensity of physical activities. This manner of determining community needs is preferable to other social or geographic criteria currently used. International norms provide a good basis for comparisons [8, 10, 15], and repetition of national surveys at intervals of perhaps 10 years will indicate secular trends. Based on such surveys, programmes can be implemented to reduce disparities between regions, and to reduce the risk of morbidity. Such programmes should begin at an early age, both in households and in community facilities.

Nutritional education will be important to overall programme success. A nutritional module should be included in the schooling of medical and paramedical personnel and teachers. Nutrition courses should also be developed at primary and secondary schools, to improve the knowledge of future parents. While it is desirable to improve the nutrition of children in developing countries, they should be protected from the pathologies associated with overnutrition and hypokinesia. The body fat content may be too high if it approaches that observed in countries where there is a high incidence of cardiovascular disease. On the other hand, high values of PWC_{170} are a desirable goal, since cardiorespiratory fitness is linked to a low risk of cardiovascular diseases [17].

Many adult pathologies have their origin during infancy; risk factors for hypertension, hyperlipidemia, and physical inactivity are frequent in the child. The prevention of coronary disease, cerebrovascular accidents, hypertension, obesity, and diabetes all begins with the learning of good health habits in childhood. A certain minimum level of physical activity is important both to the optimal development of the body, and to counteracting a later deterioration of health. The period of puberty seems the most appropriate to instil good habits, as well as being the age most vulnerable to the onset of hypokinesis. Inactivity begun at this age is difficult to correct in later life.

References

1 Morley D: Pediatric priorities in the developing world. Postgraduate pediatric series. London, Butterworths, 1974.
2 Kellou MK: Evolution des principales caractéristiques nutritionnelles de l'alimentation des algériens au cours des vingt dernières années. Alger, Institut National de Santé Publique, 1989.
3 Hadj-Lakhal B: Epidémiologie de la malnutrition protéino-calorique de l'enfant algérien d'age préscolaire. Alger, Institut National de Santé Publique, 1975.
4 Dekkar N: Croissance et developpement de l'élève algérien; thése, Université d'Alger, 1986.
5 Mac Mahon B, Pugh TF: Epidemiology. Principles and methods. Boston, Little, Brown, 1970.
6 Sempe M, Pedron C, Roy Pernot MP: Auxologie. Méthodes et séquences. Paris, Théraplix, 1979.
7 Lange Andersen K, Shephard RJ, Denolin H, Varnauskas E, Masironi R: Les épreuves d'effort. Principes fondamentaux. Genève, Organisation Mondiale de la Santé, 1971.
8 Hunsicker P, Reiff GG: Aahper Youth Fitness. Test manual. Washington, AAHPER Publications, 1976.
9 Blair NS, Falls HB, Pate RR: A new physical fitness test. Physic Sportsmed, 1983;11:87–95.
10 Organisation Mondiale de la Santé: Mesure des modifications de l'état nutritionnel. Genève, WHO, 1983.

11 Sprynar Z, Sprynarova S: Physical development of Algerian school-children. Anthropologie 1973;11:129–133.
12 Chamla MC, Demoulin F: Croissance des algériens de l'enfance à l'age adulte (région de l'Aures). Paris, Centre Nationale de Récherche Scientifique, 1976.
13 Ben Khalifa F: Caractéristiques morphologiques et biochimiques et épidémiologie du diabète dans la population de Tunis. Tunis, Imprimérie Officielle de la République Tunisienne, 1979.
14 Keller W, Fillmore CM: Prévalence de la malnutrition protéino-énergétique. Wld Hlth Statist Q 1983;36:129–169.
15 Seliger V, Bartunek Z: Mean values of various indices of physical fitness in the investigation of Czechoslovak population aged 12–55 years. Prague, CSTV, 1976.
16 Heyters C: Valeurs normales de la capacité de travail pour une fréquence cardiaque de 170 battements par minute (CT 170). Méd Sport 1985;3:38–48.
17 Pate RR: A new definition of youth fitness. Phys Sportsmed 1983;4:77–83.

Nourredine Dekkar, MD, PhD, Centre Sportif Feminin, Chemin Gaddouche, Ben Aknoun 16030, Alger (Algeria)

Physical Fitness in Children and Adolescents from Differing Socioeconomic Strata

Galo E. Narváez Pérez, Carlos P. D'Angelo, Raul D. Zabala

Laboratorio de Evaluaciones Morfofuncionales (LABEMORF), Buenos-Aires; Division Medicina del Deporte – Secretaria de Salud Publica – Municipalidad de la Ciudad de Buenos Aires; Universidad Nacional de Catamarca, San Fernando del Valle de Catamarca, Argentina

Introduction

In general, the practice of physical activity in primary and secondary schools has been based on classical models, taken from the experience of sociocultural environments much more homogenous than those found in South American countries. The lack of systematic studies showing the realities of each region has greatly accentuated the problem.

Several previous studies have demonstrated the importance of socioeconomic and nutritional factors in relation to the level of physical fitness and sports performance in children and adolescents [1–6]. The nutritional status also has a marked effect on body composition and sports performance, especially at the peak of height velocity (PHV) [7–10].

Criteria have now been established as to an appropriate starting time, intensity and duration of physical activity for school children and resultant effects on physical fitness [11, 12]. In order to consider these questions further and to suggest new solutions in a South American context, the National Plan of Physical Fitness Evaluation (NPPFE) was launched in Argentina in 1980 [13]. This program has as its original aim to diagnose fitness levels in the population between the ages of 12 and 18 years, based on a sample of 65,000 secondary school students (34,450 females and 30,550 males). These students represented a 0.5% sample of the national enrolment.

At this stage, body composition was also studied on a sub-sample living in the Federal Capital [14]. After determining tendencies for the different variables studied, the sample size was increased to 1,200,000, and a standard protocol was established, to be applied regularly at the beginning and at the end of the school-terms.

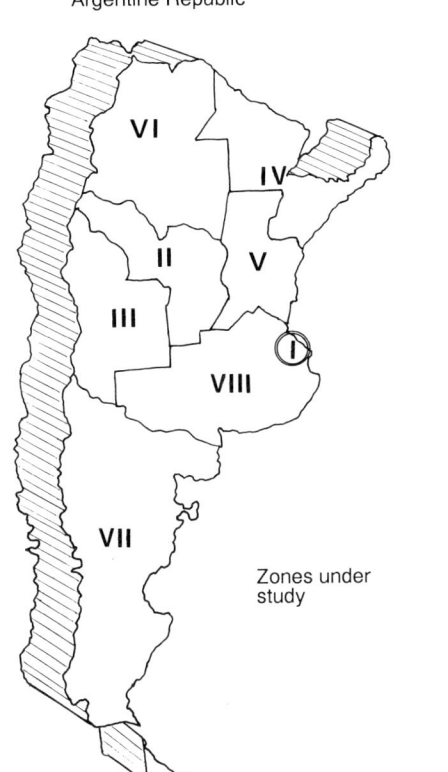

Fig. 1. Zones under study.

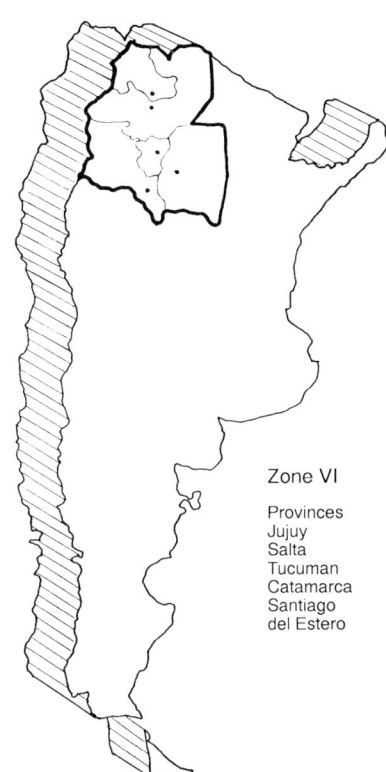

Fig. 2. Provinces within zone VI.

Zone VI

Provinces
Jujuy
Salta
Tucuman
Catamarca
Santiago del Estero

Methods

Investigators applied a battery of physical fitness tests, selected and approved by an expert committee appointed by the National Physical Education Board [13]. Seven tests were conducted: (1) a test of aerobic power (12-min run, with results expressed in ml/kg·min, using an equation described by Narváez: $\dot{V}_{O_2} = $ ln distance in 12 min × 40.4–269 [13]; (2) a test of anaerobic capacity: (40-second run); (3) a test of velocity (50-metre dash); (4) a test of flexibility (a sit-and-reach test); (5) a test of motor ability (shuttle run, SR); (6) a test of explosive power (vertical jump, VJ), and (7) at test of upper extremity power (pull-ups, PU).

Differences of fitness levels were examined for eight regions into which the country had been divided [15] (fig. 1,2). Delta percentages were established for each of the eight regions (R1 ... 8) and the averages for the whole country (WC) according to the equation:

% R1 ... 8 = [(WC−R1 .... 8)/WC] × 100.

Significance was assessed by a one-way analysis of variance (one-way Anova), and use of Duncan's Test, with acceptance of findings where $p<0.05$.

Two further independent studies were conducted to evaluate the influence of socioeconomic factors on physical fitness in primary-school children. (A) The first of these studies was conducted on 294 healthy people: 145 males and 159 females aged 9–12, drawn from two areas in the city of Buenos Aires – a residential area (RA) and a slum area (SA). (B) The second of these studies was conducted on 300 children aged 9–13, attending a private school in a residential area, all of them taking part in sports activities in addition to regular physical education classes [17].

In (A), the protocol established by the Physical Fitness Test Manual (National Physical Education Board) [13], was observed. A socioeconomic-cultural survey was also conducted, with emphasis on such aspects as the quantity and quality of sports activity, the location and type of sports undertaken, the influence of the family, the child's attitude towards sport, and physical activity in the family group [18].

To (B) were the measurement of skinfolds and some athletic tests such as: 1,000-metre run, speed over 30 metres, standing long jump, standing quintuple jump, and stick throw.

Results

The general results have appeared as an official report of the National Physical Education Board, with tables containing mean values ± 1 SD for each age group in each region and province of the Argentine Republic [19]. Tables 1 and 2 partially summarize this report. In order to make the information easy to handle, percentile tables were prepared by age and sex for each of the physical qualities under study. Findings were compared with the American population as described in the AAHPER report [20].

Differences were very significant beyond age 13 (females) and age 15 (males), with respect to both anthropometric and functional tests. However, differences of flexibility decreased after age 15 in both sexes.

It is difficult to generalize conclusions, due to the large territorial area (2,790,485 km^2, located between 21 and 55 degrees of southern latitude), to ethnic differences, to geographic and climatic differences, and to dietary habits, among other factors. In order to minimize the problem, differences in physical fitness level were tested between the eight regions into which the Argentine Republic was divided, according to criteria adopted by the educational authorities.

F values were sighificant at $p<0.001$ for the 40-second run, the VJ and the SR in the males. In the females, all tests showed very significant ($p<0.001$) regional differences. Application of Duncan's test helped to determine the differences in physical fitness and morphology between the eight regions. For example, in the 12-min run, males showed significant differences between regions 1 and 2 vs. 6-5-3-8 (table 3).

Table 1. Summary table of data for male students from 8 regions of the Argentine Republic

Age	12 years				15 years				18 years			
	body mass kg	height cm	12-min run m	shuttle run s	body mass kg	height cm	12-min run m	shuttle run s	body mass kg	height cm	12-min run m	shuttle run s
Whole country	46.9 9.1	156 9	2,299 319	12.3 1.3	59.5 8.6	168 8	2,462 306	11.8 1.2	66.1 7.9	173 7	2,477 364	11.6 1.2
Region 1	46.8 9.8	157 9	2,270 276	12.3 1.2	60.3 9	169 8	2,394 296	11.8 1.1	66.1 7.2	172 7	2,421 374	11.8 1.2
Region 2	43.4 7.7	152 7	2,321 294	12.3 1.2	57.9 9	168 8	2,533 280	11.8 1.3	66.3 7.4	174 6	2,572 357	11.3 1.2
Region 3	50.4 11.1	157 10	2,319 378	13.4 1.9	58.2 7.9	168 7	2,491 321	12.4 1.5	65.9 8.5	172 6	2,488 329	12.6 1.4
Region 4	47.8 8.7	156 8	2,242 262	12.2 1.1	58.6 8.2	168 8	2,436 313	11.7 1.3	65.7 8.4	173 7	2,467 310	11.1 1.3
Region 5	46.5 8.2	156 8	2,371 364	12.1 1.4	60.2 8.4	169 7	2,478 304	11.7 1.2	66.6 8.3	173 7	2,443 394	11.4 1.1
Region 6	44.9 8.1	154 8	2,324 332	12.1 1.3	58.1 8.7	167 8	2,486 329	11.9 1.2	63.7 7.5	171 6	2,454 376	11.4 1.1
Region 7	46.8 9.2	154 8	2,138 334	12.6 1.1	58.5 8.4	167 8	2,529 321	11.8 1.3	63.6 8.7	169 8	2,521 385	11.5 1.2
Region 8	52.1 8.1	159 9	2,287 314	12.3 0.9	60.2 8.6	169 8	2,451 299	11.8 1.2	66.9 7.9	174 6	2,478 358	11.5 1.2

Values listed are means ± SD.

Table 2. Summary table of data for female students from 8 regions of the Argentine Republic

Age	12 years					15 years					18 years			
	body mass kg	height cm	12-min run m	shuttle run s		body mass kg	height cm	12-min run m	shuttle run s		body mass kg	height cm	12-min run m	shuttle run s
Whole country	46.9	154	1,847	13.5		53.2	159	1,833	13.3		54.9	160	1,808	13.3
	7.5	6	246	1.6		7.2	6	261	1.4		6.8	6	247	1.4
Region 1	47.9	156	1,742	13.3		53.6	159	1,762	13.1		55.4	160	1,803	12.9
	5.5	8	187	1.1		7.1	6	239	1.3		7.5	6	226	1.2
Region 2	44.9	156	1,896	13.2		52.6	160	1,860	12.9		54.2	160	1,820	13.1
	7.1	6	251	1.3		6.8	6	282	1.2		5.8	6	239	1.09
Region 3	45.5	154	1,920	14.2		53.1	159	1,876	13.9		55.1	159	1,768	14.1
	8.5	6	214	1.7		6.8	5	214	1.5		7.8	6	257	1.3
Region 4	47.1	154	1,890	13.9		51.7	159	1,848	13.6		53.7	160	1,854	13.3
	7.4	6	261	1.5		6.6	6	278	1.2		6.6	6	250	1.2
Region 5	47.7	157	1,863	13.5		53.6	160	1,818	13.2		55.1	160	1,776	13.1
	8.3	7	277	1.9		7.1	6	248	1.5		6.7	6	245	1.6
Region 6	45.4	152	1,802	13.5		51.2	157	1,777	13.8		52.9	159	1,776	13.9
	6.4	6	244	1.8		7.5	6	288	1.4		6.6	6	238	1.4
Region 7	50.6	153	1,876	12.9		54.1	159	1,844	13.3		54.2	159	1,885	13.7
	3.9	5	194	0.5		7.4	5	208	1.3		7.2	7	246	1.3
Region 8	49.3	155	1,822	13.3		53.6	159	1,867	13.2		55.6	160	1,823	13.4
	8.3	7	230	1.4		7.2	6	259	1.3		6.5	5	255	1.4

Values listed are means ± SD.

Table 3. Delta values for male subjects from 8 regions of Argentine Republic, with application of Duncan's Multiple Range Test

Body mass Region	8	1	5	3	4	7	2	6
% Delta	−2.75	−0.91	−0.59	0.38	0.63	1.16	2.62	3.77

F = 5.64

12-min run Region	2	7	6	5	3	8	4	1
% Delta	−2.57	−0.94	−0.65	−0.65	−0.16	0.17	0.76	2.19

F = 5.44

40-second run Region	4	6	3	2	5	8	7	1
% Delta	−2.02	−1.18	−0.67	−0.37	−0.05	0.38	0.92	1.6

F = 12.38

Shuttle run Region	3	1	7	8	6	2	5	4
% Delta	−6.72	−0.09	0	0.28	0.41	0.71	1.09	2.04

F = 50.90

Differences between values underlined are significant at $p<0.05$.
Delta percentages are expressed for the values of each in the eight regions (R1 ... 8) and the average for the whole country (WC), according to the following equation:

$$\% \text{ deltas} = R1 \ldots 8 = (WC - R1 \ldots 8)/WC \times 100.$$

Comparing data with the US population, our population was located at the 10th percentile in the shuttle run, at the 55th percentile for pull-ups at age 13, and (contrary to what might be expected) percentiles progressively decreased with age, to reach the 30th percentile at age 17 years.

The 12-min run presented special characteristics. Although males were at the 5th percentile at age 13, they quickly moved to the 20th percentile at

Table 4. Morphofunctional values for males

Age	9 years		10 years		11 years		12 years	
	Fatima	S. Gaynor	Fatima	S. Gaynor	Fatima	S. Gaynor	Fatima	S. Gaynor
Body mass, kg								
X̄	29.49	35.78	31.27	35.81	36.46	39.47	38.39	45.40
SD	5.96	6.06	4.48	5.42	6.64	7.03	6.64	7.52
t	2.27*		2.59**		1.37		2.68**	
n	10	9	25	19	25	19	10	9
Height, cm								
X̄	129.10	135.78	135.52	141.31	141.79	143.13	145.74	151.21
SD	4.25	7.61	6.72	3.97	4.89	7.16	6.43	5.93
t	2.325**		3.33**		0.65**		2.77**	
Trunk length, cm								
X̄	64.30	69.78	68.80	72.90	70.90	74.93	74.42	76.42
SD	3.94	4.71	4.33	2.44	3.20	3.33	4.44	4.50
t	2.73**		3.73**		3.865**		1.371	
Lower limb length								
X̄	64.80	66.00	66.72	68.31	70.92	68.20	71.32	74.79
SD	3.19	4.42	4.69	2.32	4.12	5.31	5.44	3.47
t	0.67		1.39		1.78		2.42**	
50-meter dash, s								
X̄	10.95		10.02	9.10	9.09	8.94	9.13	8.61
SD	0.37		1.62	0.38	1.11	0.35	0.99	0.67
t			2.69**		0.66		1.75	
40-second run, s								
X̄	217.92	146.50	176.00	217.92	203.93	211.47	211.89	221.32
SD	28.78		37.88	14.31	36.82	12.34	36.10	17.99

Variable											
kg/s											
X̄	108.68		138.92		186.19		208.88		204.91		251.02
SD	35.03		41.54		47.47		39.33		56.58		45.62
t		1.64		4.897**		0.999		0.98		4.76**	
12-min run, m											
X̄	1,896.8		1,935.7		2,187.6		2,219.7		2,225.8		2,303.4
SD	282.8		398.9		260.6		225.6		333.7		264.1
t				4.76**		1.687				4.76**	
V̇O₂, litres/min											
X̄	1,047.3		1,126.2		1,504.2		1,651.4		1,594.2		1,965.7
SD	304.0		227.5		297.1		285.6		320.9		344.4
t				3.21**		0.424		0.68		3.00**	
V̇O₂, ml/kg·min											
X̄	35.40		36.19		41.33		42.01		41.84		43.40
SD	6.25		6.30		4.77		3.94		6.30		4.47
t				4.53		1.54		0.46		0.76	
Vertical jump, cm											
X̄	21.16		20.56		23.03		27.65		23.42		
SD	4.67		4.38		6.12		3.70		4.74		
t				3.24**				2.53*			
kg/s											
X̄	30.28		31.28		38.25		47.96		40.94		
SD	8.60		6.00		7.94		7.93		8.37		
t								15.9*			
Shuttle run, s											
X̄	14.83		13.83		13.58		13.62		12.69		13.43	12.05
SD	1.69		0.37		1.44		1.95		0.89		2.00	1.00
t				0.31		1.72				2.76**	

School Fatima is representative of a slum-type area (SA); school S. Gaynor is representative of a residential area (RA). Significantly different: *$p<0.05$; **$p<0.01$.

Table 5. Morphofunctional values for females

	Age											
	9 years			10 years			11 years			12 years		
	Fatima	S. Gaynor	t	Fatima	S. Gaynor	t	Fatima	S. Gaynor	t	Fatima	S. Gaynor	t
Body mass, kg												
x̄	43.43	42.73	0.32	38.34	39.71	0.52	32.06	34.64	1.17	30.64	31.79	0.49
SD	8.59	7.28		6.90	8.76		6.32	6.27		5.14	5.93	
n	22	22		33	14		17	24		18	9	
Height, cm												
x̄	148.23	151.23	1.71	141.10	143.71	1.53	134.59	138.18	0.99	130.55	134.94	1.75
SD	6.45	5.35		5.25	9.39		7.38	6.27		5.88	7.01	
Trunk length, cm												
x̄	76.45	77.45	0.94	71.60	74.50	1.93	67.94	72.35	3.31**	65.22	70.78	3.89**
SD	3.83	3.77		3.14	5.21		4.52	3.12		3.31	3.84	
Lower limb length, cm												
x̄	71.77	73.77	1.38	69.45	69.21	0.14	66.65	65.82	0.48	65.33	64.17	0.73
SD	5.81	3.74		3.86	5.63		5.96	3.81		3.97	3.79	
50-metre dash, s												
x̄	10.05	9.11	2.22*	10.79	9.56	3.63**	11.50	9.82	5.34*	11.17		
SD	5.81	0.60		1.63	0.69		1.17	0.56		1.14		
40-second run, m												
x̄	188.00	205.35		175.88	202.75		156.35	197.82		151.22		
SD	28.25	23.45		29.04	10.94		31.65	11.07		19.32		

		t				t				t		
kg/s												
X̄	203.83	2.01**	219.01	167.81	4.6**	203.36	127.12	5.09**	171.25	114.64		
SD	51.06		36.34	36.40		41.37	44.22		33.18	14.64		
t		3.39**			2.79**			3.39**				
12-min run, m												
X̄	1,751.5		1,848.0	1,846.8		1,809.9	1,755.2		1,765.0	1,815.6		1,962.2
SD	318.0		182.5	264.3		228.9	269.4		255.1	212.5		470.0
t		1.07			0.48			0.11			1.11	
V̇O₂ litres/min												
X̄	1,380.9		1,450.6	1,293.9		1,375.7	1,033.7		1,154.9	1,031.8		1,133.5
SD	410.4		213.3	214.3		317.8	283.7		282.1	195.3		415.7
t		0.61			0.88			1.35			0.86	
V̇O₂ ml/kg·min												
X̄	31.94		34.57	34.35		33.59	32.26		34.27	33.82		35.24
SD	7.38		4.01	5.98		5.63	6.31		7.61	4.99		10.48
t		1.24			0.42			0.92			1.14	
Vertical jump, cm												
X̄	23.01		31.44	20.97		27.73	18.76					
SD	4.94		3.72	5.35		4.86	3.51					
t		5.99**			4.23**							
kg/s												
X̄	46.18		54.23	38.39		45.04	30.64					
SD	11.71		9.40	7.01		10.20	6.72					
t		2.59**			2.22**							
Shuttle run, s												
X̄	13.25		13.17	14.53		13.28	15.53		14.43	14.33		14.58
SD	2.58		1.43	2.44		0.78	2.96		1.25	1.79		1.36
t		0.14			2.64			1.44			0.36	

School Fatima is representative of a slum-type area (SA); school S. Gaynor is representative of a residential area (RA). Significantly different: *p<0.05; **p<0.01.

age 14, to the 55th percentile at age 15, to the 65th at age 16 and to the 85–90th percentile at age 17. In the case of females, they started at the 10–15th percentile at age 13 and they remained at the 20th percentile from 14 up to 17 years of age.

The same tendency was observed with respect to abdominal muscle power, our population being located at the 30–35th percentiles.

Significant differences ($p<0.05$) were found between those from RA vs SA; in the case of males, this was true of both anthropometric and functional tests, but in females, only the functional tests differed significantly (tables 4, 5).

The socioeconomic-cultural survey showed significant differences ($p<0.01$) regarding various factors, including the quality and the quantity of sports activity, the place and type of sports, the starting age for sports, the influence of the family group, and the limited participation of females in sports. Children from RA (as opposed to those from SA) shared the following characteristics:

(1) Greater sports activity ($\chi^2 = 44.3$, $p<0.001$).
(2) The practice of sports requiring a suitable infrastructure and investment in sports equipment.
(3) Parents showing a cooperative attitude ($\chi^2 = 39.2$, $p<0.001$).
(4) Practice of sport by the father ($\chi^2 = 42.6$, $p<0.001$) and the mother ($\chi^2 = 59$, $p<0.001$).
(5) Earlier and more sustained involvement in sports competition ($\chi^2 = 27$, $p<0.0001$).
(6) Attendance at sports events with their family.

In experiment (B), it was observed that:

(1) The velocity in the 50-metre dash was the same as that in the residential area from study A (fig. 3).
(2) The velocity in the 50-metre dash was significantly dependent ($r^2 = 0.58$) on the aerobic endurance (1,000-metre run), on the explosive muscular power (SLJ) and, to a lesser degree, on the age, according to the following equation:

50-metre dash = 11.76 + (1,000-metre run time × 0.80) − (SLJ × −0.79) − (age × −0.20).

(3) The velocity curve in a 50-metre dash tends to respond to processes of biologic maturation, such as chronologic age, body mass, fat percentage, and height. However, these factors are not independent (fig. 4–6).

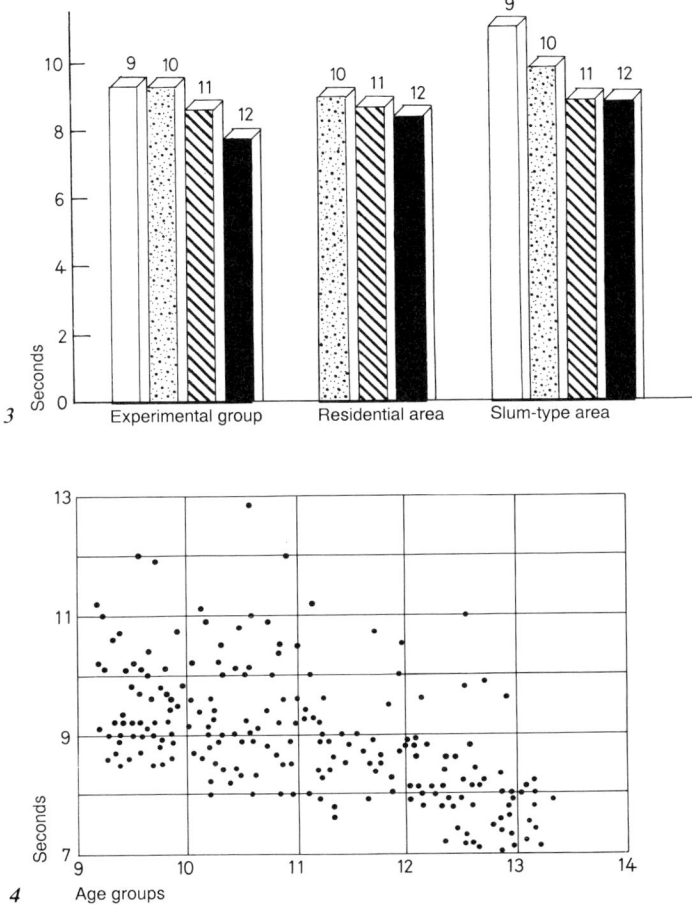

Fig. 3. A comparison of times for 50-metre dash between experimental group and students from residential and slum areas (male subjects).
Fig. 4. The influence of age on times for 50-metre dash (male subjects, n = 230).

Discussion

One of the first points to be considered is the tendency of females to keep a constant physical fitness level from menarche to ages 17–18. Although the results found are similar to those reported for other populations, there is no basis for believing that physical capacity, aerobic power, muscular power, and flexibility could not be developed further. It would therefore be a mistake to accept as inevitable the results shown in figures 7–9, where

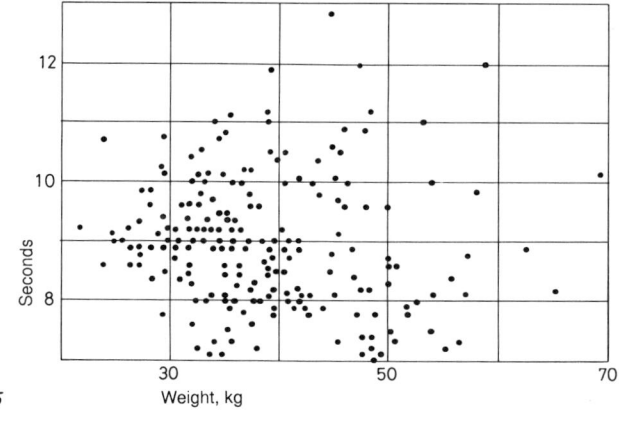

5

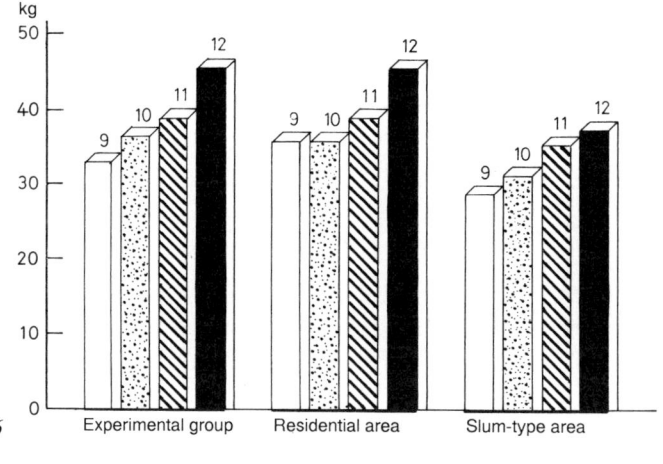

6

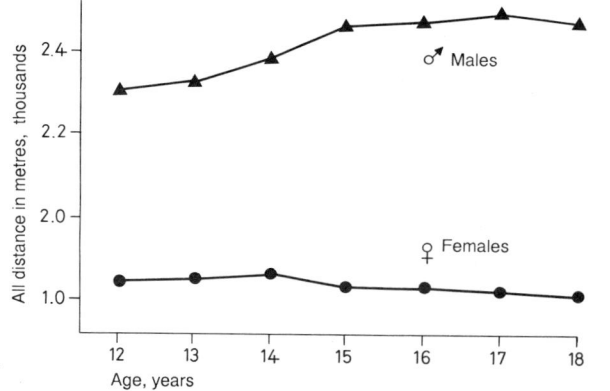

7

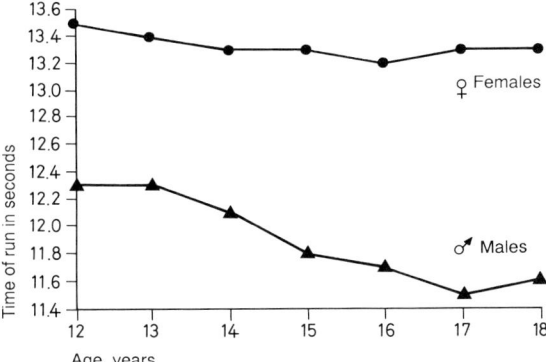

Fig. 8. Influence of age on shuttle run scores. Average for whole country.

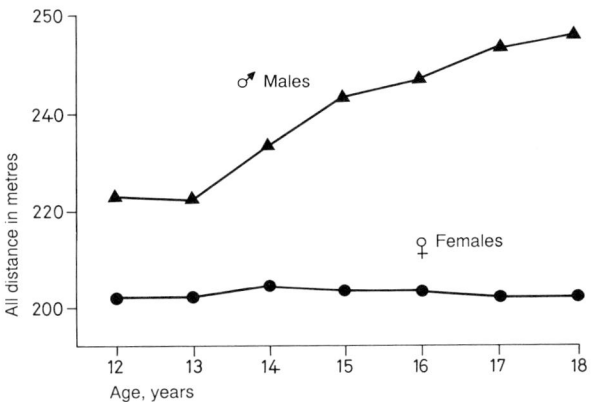

Fig. 9. Influence of age on 40-second running distance. Average for whole country.

Fig. 5. The influence of body mass on times for 50-metre dash (male subjects, n = 230).
Fig. 6. A comparison of body mass between experimental group and students from residential and slum areas (males).
Fig. 7. Influence of age on distance covered in 12-min run. Average for whole country.

all fitness values decreased between the ages of 13 and 17 years, in females more markedly than in males. This not only affects sports results adversely, but also harms the health development of the future adult population.

It is very common to see programs promoting physical activity for populations between 30 and 50 years of age. Would it not be more convenient to modify programs and teaching methods to ensure that physical condition is optimized during pre-puberty and puberty?

Determination of physical fitness in the child and adolescent population not only reveals current fitness levels, but also allows the scientist to develop new procedures, with a view to their application in teaching institutions, and to correcting the limitations of traditional methods.

An experimental study conducted on a sample of the same population could well serve as reference. An aerobic training program [21] was applied for a period of 6 weeks. The increase in oxygen uptake was statistically larger than that of a control group of similar characteristics who had received only regular physical education lessons.

From the complementary studies A and B, we may conclude that socioeconomic factors greatly influence the development of physical fitness in children and adolescents; and that the differences found between the two socioeconomic strata justify a deep cause and effect analysis at an early age, or even for the ontogenic periods, as suggested by Parizková [22, 23]. Nevertheless, fitness deficiencies in our populations are not due simply to socioeconomic influences.

The comparative study of the 8 regions of the Argentine Republic [15] revealed a certain homogeneity. Differences of morphological and structural factors (tables 1, 2) were observed, especially in regions where ethnic characteristics tended to be more homogeneous. This was the case for region VI, where the percentage of racially mixed students was proportionally greater than that of European immigrants.

Tables 3 and 6 show the superposition of variances for several physical qualities. This suggests the interaction of a number of socioeconomic, environmental and cultural factors that are characteristic of each region. It is very difficult to isolate the influence of these factors in free-living populations. Therefore, the results obtained from studies A and B cannot be extrapolated or applied in a general sense to all of the Argentinian population. However, certain partial conclusions can be attempted.

The results of D'Angelo et al. [16] show that the differences between the two groups under study are self-evident, as far as morphological and functional aspects are concerned. The highest level of physical fitness is reached by the group from the residential area (urban area). This is in accordance with the results of our socioeconomic survey [18] and the earlier work of Renson et al. [6].

Table 6. Delta values for female subjects from 8 regions of Argentine Republic, with application of Duncan's multiple range test

Body mass Region	8	1	7	5	3	4	2	6
% Delta	−1.68	−1.23	−1.23	−0.89	0.89	1.55	1.7	3.62

$F = 11.62$

12-min run Region	2	7	3	4	8	5	6	1
% Delta	−1.43	−1.39	−1.08	−1.05	−0.79	0.3	1.72	3.01

$F = 10.38$

40-second run Region	5	8	7	3	1	2	4	6
% Delta	−3.07	−1.6	−0.2	0.15	0.18	1.44	2.68	4.48

$F = 8.20$

Shuttle run Region	3	6	4	8	5	7	1	2
% Delta	−5.26	−3.2	1.62	0.67	1.12	1.25	1.59	2.17

$F = 36.31$

Differences between values underlined are significant at $p < 0.05$.
Delta percentages are expressed for the values of each of the eight regions (R1 ... 8) and the average for the whole country (WC), according with the following equation:

$$\% \text{ Deltas} = R1 \ldots 8 = (WC - R1 \ldots 8)/WC \times 100$$

In the work presented by other authors [1, 2, 5], conversely, the highest levels of physical fitness have been found in groups from rural areas. These differences may reflect, among other causes, early nutritional factors, as reported in long-term studies [1, 2, 5, 6]. Both population groups show that body mass and height differentiate definite age groups, which suggests that in both samples perfect adaptations of the factors that determine biological maturation have been achieved. However, it is possi-

Table 7. Age groups with significant differences

	Males		Females	
	S. Gaynor	Fatima	S. Gaynor	Fatima
Body mass				
	12-10-9	12-10-9	12-10-9	12-10-9
Height				
	12-10-9	12-10-9	12-10-9	12-11-10-9
Lower limb length				
	12-11-10-9	12-9	12-10-9	12-10-9*
Trunk length				
	12-10-9	12-11-10-9	12-10-9	12-11-10-9
50-metre dash				
	12-10	12-9		12-10*
40-second run				
	12-11-10	12-9		12-10-9*
$\dot{V}_{O_2}$, litres/min				
	12-9	12-10-9	12-10-9*	12-10
Shuttle run				
	12-10	n.s.	n.s.	n.s.

The differences were established between age groups through ANOVA and the Scheffe multiple range test was used to locate significant differences among mean values. For all of the Scheffe tests, the significance level was $p<0.001$.
Asterisks indicate that $p<0.05$.

ble that in the rural group (SA), the practice and guided learning of motor abilities has more strictly respected the stages of growth and maturation, as demonstrated in the report by Blischke and Quell [24].

Although the above statement should only be regarded as speculative, the ANOVA for the shuttle run (table 7) shows insignificant differences for the SA group, but allows us to distinguish, among the RA group, two age groups of males aged 10 and 12 respectively. This agrees with the work of Zabala and Narváez [17], where there is a total coincidence of results with respect to the RA group of the 'A' study. Here, velocity is measured by the 50-metre dash, and 58% of the variance is explained by the interaction of aerobic power, explosive muscular power and age. If these factors are in turn affected by the socioeconomic level to a significant degree, there should be no doubt that we are faced by a problem requiring new comparative studies and long-term observations.

In the meantime, it is important to consider the results of the socioeconomic survey conducted by D'Angelo et al. [18], which demonstrates that, apart from the morphological and functional factors discussed above, fitness results are affected partly by controllable factors. These factors include the quantity and quality of sports activity, the sports infrastructure, the parents' cooperative attitude, the father and mother's sports practice and the starting age for sports.

References

1 Harsha DW, Voors AW, Berenson GS: Racial differences in subsutaneous fat patterns in children aged 7–15 years. Am. J. Phys. Anthropol 1980;53:333–337.
2 Malina RM: Physical activity, growth, and functional capacity; in Johnson FE (ed): Physical Growth and Maturation. New York, Plenum Press, 1980, pp 303–327.
3 Narváez PGE: Distribución de grasa corporal. Buenos Aires, Associación Médica Argentina, Sociedad Argentina de Nutrición, 1982.
4 Pařízková J: The impact of ecological factors and physical activity on the somatic and motor development of preschool children; in Shephard RJ (ed): Physical Fitness Assessment Principles, Practice and Application. Springfield, Thomas, 1978, pp 238–247.
5 Pařízková J: Effect of increased physical activity on body composition during growth in different groups of children: Longitudinal studies; in Fat and Physical Fitness. Prague, Avicenum, 1977, pp 118–168.
6 Renson R, Beunen G, De Witte L, Ostyn M, Simons J, Van Gerven D: The social spectrum of the physical fitness of 12 to 19 year old boys; in Ostyn M (ed): Kinanthropometry II. Baltimore, University Park Press, 1980, pp 104–118.
7 Pařízková J: Role of body dimensions and body composition. Physical fitness and performance during growth and adulthood. Coll Antropol 1979;3:49–58.
8 Pařízková J: Impact of age, diet and exercise on man's body composition. Ann NY Acad Sci 1963;110:661–673.
9 Pařízková J: Nutritional status, somatic and functional development in preschool children as related to ecological factors and exercise. Act Facult Med Univ Brunensis 1976;57:333–340.
10 Tanner JM, Whitehouse RH, Takaishi N: Arch Dis Childh 1966;41:454.
11 Shephard RJ, Lavallée H, Jéquier JC, Rajic M, LaBarre R: Additional physical education in primary school: A preliminary analysis of the Trois-Rivières regional experiment. Baltimore, University Park Press, 1978, pp 306–316.
12 Shephard RJ: Programs of physical activity for the primary school child – needless or a necessity? In Burke E (ed): Exercise, Science and Fitness, Ithaca, Mouvement Publications, 1977, pp 70–78.
13 Narváez PGE, Echavarria A, Gonzalez L, Ledesma F, Marty B, Masabeu E, Melo H, Occhi M, Rada B, Rolandelli A, Rozada B, Zorzenón J: Manual de tests de aptitud fisica; in Dirección Nacional de Educación Fisica Deportes y Recreación (ed): Plan Nacional de Evaluación de la Aptitud Fisica. Buenos Aires, Dirección Nacional de Educación Fisica, 1980, pp 23–59.
14 Narváez PGE: Estudio de la composición corporal de una muestra de poblición juvenil Argentina; in Memorias de la XXVIII Reunión Cientifica Sociedad Argentina de Investigacián Clinica (ed): XXVIII Reunión Cientifica, Mar del Plata, 1983, p 309.
15 Narváez PGE: Physical fitness in different regions of Argentine Republic. Ann 12th Int Congr Anthropological and Ethnological Sciences, Zagreb, 1988.

16 D'Angelo C, Narváez PGE, Sgala B, Materola A: El nivel de aptitud fisica en estudiantes de escuela primaria de distintos niveles socioeconómicos. Medicina del Ejercicio 1987;1:2–6.
17 Zabala DR, Narváez PGE: Medición de la aptitud fisica en niños: Criterios para la evaluación 1990. Montevideo, Medicina del Ejercicio, accepted.
18 D'Angelo C, Narváez PGE, Sgala B, Materola A: Estudio de factores socio-culturales para la prática de deportes en el niño. Med Ejercicio 1987;1:11–17.
19 Narváez PGE, Echavarria A, Gonzalez L, Ledesma F, Marty B, Masabeu E, Melo H, Occhi M, Rada B, Rozada B, Zorzenón J: Informe sobre los resultados de las mediciones antropométricas y las pruebas de aptitud fisica: Conclusiones; in Dirección Nacional de Educación Fisica Deportes y Recreación (ed): Plan Nacional de Evaluación de la Aptitud Fisica. Buenos Aires, Dirección Naciónal de Educación Fisica, 1984.
20 American Alliance for Health, Physical Education, and Recreation: Comparing test results; in Hunsicker (ed): AAHPER Youth Fitness Test Manual. Washington, AAHPER Publications, revised edn, 1976, pp 37–53.
21 Narváez PGE, Alvarez CJJ, Zabala R, Barbieri C: Entrenamiento de la potencia aeróbica: Cualidad Resistencia Cardio-Respiratoria; in Labemorf (ed): Manual de Evaluación y Entrenamiento. Buenos Aires, Publicaciones Labemorf, 1984, pp 36–44.
22 Pařízková J: The impact of nutritional factors during early ontogeny; in Roche AF (ed): Nutrition and Malnutrition. New York, Plenum Press, 1974, vol 49, pp 120–149.
23 Pařízková J: Nutrition and work performance; in Chandra RK (ed): Critical Reviews in Tropical Medicine. Boston, Massachusetts Institute of Technology, 1970, vol 1, pp 307–328.
24 Blishke K, Quell M: Socially determined variation In sensoriomotor behavior; in Ostyn M (ed): Kinanthropometry II. Baltimore, University Park Press, 1980, pp 104–118.

Galo E. Narváez Pérez, MD, Laboratorio de Evaluaciones, Morfofuncionales (LABEMORF), Amenabar 783 (1426), Buenos Aires (Argentina)

Aerobic and Anaerobic Physical Capacity of Cuban Schoolchildren Subjected to Different Motor Regimens

Nelson Arbesú

Institute of Nutrition, Laboratory of Physiology, Havana, Cuba

Introduction

Assessment of the nutritional and health status of a population requires a knowledge of the physical capacity of the inhabitants. The assessment of physical capacity in children and adolescents is particularly necessary in most countries nowadays, due to the negative influence of hypokinesis in these age groups. An excess of physical activity may also be harmful, particularly when the body is not ready to withstand an intensive motor regimen. In order to optimize the effectiveness of school physical education programs it is necessary to provide an objective physiological assessment of the different motor regimens adopted in various schools. Moreover, although there is a large amount of experimental data on the functional capacity of children and adolescents in developed countries, little is known about the new generations of developing countries particularly in Latin America.

Materials and Methods

The present paper describes three series of investigations made on a total of 635 male subjects, ranging from 10 to 16 years of age. Of the total, 100 belonged to *special high schools*, where students combined the teaching activities of their grade with physical agricultural tasks for half of each day. Cuba has a national plan called *'high schools in the countryside'* that includes hundreds of schools with a boarding system. According to Serralta [1], the required agricultural physical activities in such schools induce an average heart rate of 117 beats/min (maximum 136, minimum 107). The experimental data collected at such institutions were compared with those found in two other schools, a *special sports school* (n = 100) where students trained for 2.5–3 h a day, 5 times a week, and an urban school (n = 100) where students only met the minimum physical education program required by our national system of education. Physical performance capacity was measured in each case by the PWC_{170}

method. The basic indices of physical development, body mass, and height were also estimated.

We subsequently investigated a second series of 182 children and adolescents from another sports school, where subjects had been grouped into representatives of *team games* (basketball and baseball) and *rhythmic sports* (swimming and track-and-field).

A third series of 50 schoolchildren were studied longitudinally, twice a year. They belonged to a *Special Sports Center for Athletic Development*. Such students undertook a special training program without a *close specialization* 3 h per day. In addition, 103 children from a normal *sports school* (control group) were investigated.

In the first series of experiments, the physical work capacity was estimated by the method of Wahlund [2], adjusting the power output at each test stage to the characteristics of our schoolchildren. A Monark mechanically braked cycle ergometer was used for the PWC_{170} determinations. The heart rate was recorded with a Sharp electrocardiograph, using Nehb's technique for the trunk pick-up points. Measurements were taken on-site at the different schools. Measurements on subjects from the second and third series included skinfold thickness, height, body mass, $\dot{V}_{O_2}$ max and the parameters derived therefrom. The anaerobic threshold and the oxygen debt plus its fractions were also studied, measurements being taken at the beginning and at the end of the school year. Data in the second and the third series of experiments were collected in the Physiology Laboratory at the Institute of Physical Education in Havana. A Dutch-made Mijnhard spiroergometer was used, with a stepwise increment of loadings. The open-circuit gas analyzer gave minute-by-minute values for respiratory volume ($\dot{V}_E$, ATPS, STPD, and BTPS), oxygen consumption ($\dot{V}_{O_2}$ STPD), heart rate, respiratory frequency, and respiratory gas exchange ratio.

The power output (W) was adjusted to cause a similar biologic response in all subjects (table 1), the initial power output being based on our cumulative experience, together with observation of the individuals's body mass and state of biological development. In order to determine the second power output (N_2), the first was multiplied by two. Then, the power output was raised by the equivalent of the first load (N_1) every 3 min. If the heart rate at the first stage was lower than 100 beats/min, or higher than 130 beats/min, then the power was raised or lowered by 5 or 10 W, respectively, table 2 shows the actual values of heart rate and power output (W) in a group of 25 eleven-year-old sports children. The biologic response was very similar to the intended values.

During the last minute of the test, we asked for a supramaximal effort, and oxygen debt repayment was followed for 30 min following the test. The alactate and lactate components of the debt were estimated using an analytical method developed in our laboratory [3]. Anaerobic threshold was estimated on the basis of excess CO_2 [4].

Results and Discussion

The physical characteristics of the subjects taking part in the first series of experiments are shown in table 3. Data are for an urban high school, where 44 h of physical education are taught each semester; another *school in the countryside*, where agricultural tasks are performed for part of each day (for instance, collecting fruits and vegetables), and a *special sports school* where students practice swimming everyday. Height and body mass indices are very similar for the *school in the countryside* and the urban school. However, students attending the *sports boarding school* excel

Table 1. Average heart rate at each work stage (N_1-N_5)

	Heart rate, beats/min				
	110–120	130–140	150–160	170–180	185–200
W	N_1	N_2	N_3	N_4	N_4

Table 2. Actual heart rate and power output during $\dot{V}_{O_2}$ max determination in 11-year-old children at a sports school (mean ± SD).

	N_1	N_2	N_3	N_4	N_5
Heart rate, beats/min	119 ± 6	142 ± 7	159 ± 8	176 ± 12	187 ± 10
Power Output, W	22 ± 2	43 ± 2	61 ± 3	80 ± 4	115 ± 8

Table 3. Physical characteristics of subjects participating in the first series of experiments (mean ± SD)

Age years	Urban high school (n = 100)		School in the countryside (n = 100)		Sport boarding school (n = 100)	
	height cm	body mass kg	height cm	body mass kg	height cm	body mass kg
11–12	144.6 ± 6	35 ± 5	143 ± 8	27.6 ± 6	150 ± 5	38.5 ± 5
13	150 ± 8	37 ± 5	150 ± 6	40 ± 9	160 ± 7	47.5 ± 5
14	157 ± 5	45 ± 6	156 ± 8	45 ± 9	163 ± 7	52 ± 6
15–16	164 ± 6	52 ± 8	164 ± 8	55 ± 11	171 ± 6	60 ± 3

over the other students with respect to both of these variables ($p < 0.05$), probably because body build is a criterion in selection for these schools. Children entering sports schools are close to or above the 90th percentile of height for the Cuban population. The group aged 11–12 years are about 7 cm taller than their peers, but there is less difference of body mass. As the children become older, height differences become increasingly discrete, while they also develop for body mass. Suspecting inter-group differences of physiological maturity, we assessed sexual development according to Tanner's [5] method, for the axilla and the pubes (5-point score system) (table 4). Plainly, the sports children show accelerated sexual development as compared with other children of similar age. It is unclear whether

Table 4. Degree of sexual (physiologic) maturity in the three groups of subjects: percentages are based on Tanner's [5] schema

Age years	Urban high school (n = 100)	School in the countryside (n = 100)	Sports boarding school (n = 100)
11–12	A0 P0–91%	A0 P0–94%	A0 P0–61%
	A0 P1–9%	A0 P1–6%	A0 P1–39%
13	A0 P0–46%	A0 P0–48%	A0 P0–1%
	A0 P1–54%	A0 P1–52%	A0 P1–60%
14	A1 P2–90%	A1 P2–97%	A2 P2–56%
	A2 P3–10%	A2 P3–3%	A2 P3–44%
15–16	A1 P2–22%	A1 P2–29%	A2 P3–4%
	A2 P3–78%	A2 P3–71%	A2 P4–96%

Table 5. Physical work capacity (PWC_{170} of subjects, mean $\pm$ SD): data are expressed in kpm/min; 6 kpm/min approximates 1 W

Age years	Urban high school (n = 100)		School in the countryside (n = 100)		Sports boarding school (n = 100)	
	PWC_{170}	PWC_{170}/kg	PWC_{170}	PWC_{170}/kg	PWC_{170}	PWC_{170}/kg
11–12	435 ± 46	12 ± 2	557 ± 57	14 ± 2	515 ± 55	14 ± 2
13	530 ± 72	14 ± 2	584 ± 62	15 ± 2	620 ± 66	14 ± 3
14	525 ± 58	11 ± 2	604 ± 62	13 ± 2	780 ± 65	15 ± 2
15–16	691 ± 84	13 ± 2	816 ± 73	16 ± 2	1,023 ± 99	18 ± 3

the difference is secondary to genetic factors, or is stimulated by the systematic programme of physical exercise. The higher degree of physical maturity coupled with the impact of systematic physical training is reflected in the indices of physical work capacity (table 5). Except for the 11- to 12-year-old group, children attending the sports school greatly excel over their age-matched peers. The influence of systematic training is probably not yet apparent at 11–12 years. Comparisons between the *school in the countryside* and the urban school are of particular interest because there are no significant differences of physical development between these two groups. However, PWC_{170} values are much higher at the *school in the countryside* than in urban school students. The physical activities required at the rural school thus favor the development of aerobic power. Between

13 and 14 years of age, a sharp increase in body mass occurs in both groups, 8 kg in students from the urban school and 5 kg in students from the *school in the countryside*. The first group has an associated decrease of PWC_{170}/kg of 21%, while in students from the *school in the countryside* the decrease is only of 13%. These results suggest that school-based agricultural physical activities cause a better functional cardiovascular adjustment to the somatic changes of puberty. In children attending the sports school, we recorded a still better adjustment of functions to this process.

In another series of experiments we investigated sport students aged 12–13 years. They were stratified according to their chronological age and the number of years of sport training that they had undertaken. The population that was screened had the following number of years of *close specialization*: 12 year olds, 1 year; 13 year olds, 2 years; 14 year olds, 3 years; 15 year olds, 4 years. Table 6 shows that the body mass and the height of students trained in *team games* is higher than that of those trained in *rhythmic sports*. These differences presumably reflect selection criteria for entry into the two specialities. The analysis of sexual development does not show significant intergroup differences.

Both the absolute peak power in kpm/min (W_{max}) and the $\dot{V}_{E\,max}$ (BTPS) of schoolchildren practicing *team games* show higher values at all ages than the group practicing *rhythmic sports*. Only after 4 years of systematic training (at 15 years of age) do values for the two groups start to be alike. The same may be concluded when analyzing the absolute values of $\dot{V}_{O_2}$ max (STPD) (tables 7, 8).

The data thus show a significant difference of aerobic power favoring children who practice *team games*. The $\dot{V}_{O_2}$ max/kg suggests that at these ages, the motor regimen requiring *team games* causes a marked cardiovascular and respiratory development that is related to growth in the morphological structure of the body.

According to Shephard [6], Macek and Vavra [7], and Raczek and Brehmer [8], the relative AT is very high in children, reaching 80–85% of the $\dot{V}_{O_2}$ max. Our values for schoolchildren from Havana City differ markedly from those of other authors (table 9). In our samples, the anaerobic threshold at all ages is around 60% of the $\dot{V}_{O_2}$ max.

The third sequence of trials shows the effectiveness of a system of multilateral training. This is put into practice before starting *close sports specialization*. The students attending special centers practice a group of sports, simultaneously and without *competitive stress*, during all of the school year. Measurements were made in August–September, before the beginning of the course, and in May–June. At the time of the second measurement, we included a control group from sports schools in Havana, where the students practice only one sport (*close specialization*). The

Table 6. Physical characteristics of students who practiced team games and rhythmic events (mean ± SD)

Group	Team games	Rhythmic events
12 years (n = 38)		
Height, cm	160 ± 5	150 ± 4.3
Body mass, kg	51.3 ± 6	37.6 ± 4
13 years (n = 54)		
Height, cm	165.7 ± 9	156 ± 7.2
Body mass, kg	56 ± 7	48.3 ± 6
14 years (n = 53)		
Height, cm	170 ± 8.3	162.6 ± 8
Body mass, kg	63 ± 6	50.8 ± 8
15 years (n = 37)		
Height, cm	168 ± 9	166 ± 9
Body mass, kg	61.3 ± 7	58.2 ± 6.4

Table 7. Peak power output and $\dot{V}_{E\,max}$ (BTPS) of the subjects who practice team games and rhythmic events (mean ± SD): the power output is expressed in kpm/min (6 kpm/min approximates 1 W)

Group	Team games	Rhythmic events
12 years (n = 38)		
W max, kpm/min	892 ± 209	736 ± 96
$\dot{V}_{E\,max}$, liters/min	91 ± 19	53.8 ± 7
13 years (n = 54)		
W max, kpm/min	1,032 ± 234	762 ± 106
$\dot{V}_{E\,max}$, liters/min	100 ± 18	60 ± 8.2
14 years (n = 53)		
W max, kpm/min	1,152 ± 243	862 ± 187
$\dot{V}_{E\,max}$, liters/min	101 ± 23	66.8 ± 7.8
15 years (n = 37)		
W max, kpm/min	1,214 ± 206	1,020 ± 190
$\dot{V}_{E\,max}$, liters/min	102 ± 23	73.5 ± 11

Table 8. $\dot{V}_{O_2}$ max (STPD) and $\dot{V}_{O_2}$ max/kg of subjects who practice team games and rhythmic events (Mean ± SD)

Group	Team games	Rhythmic events
12 years (n = 38)		
$\dot{V}_{O_2}$ max, ml/min	2,451 ± 445	1,650 ± 334
$\dot{V}_{O_2}$ max (ml/kg · min)	47.6 ± 5	42.3 ± 3.9
13 years (n = 54)		
$\dot{V}_{O_2}$ max, ml/min	2,918 ± 384	2,038 ± 377
$\dot{V}_{O_2}$ max (ml/kg · min)	52 ± 4.3	42.9 ± 5.4
14 years (n = 53)		
$\dot{V}_{O_2}$ max, ml/min	3,192 ± 486	2,224 ± 439
$\dot{V}_{O_2}$ max (ml/kg · min)	50.2 ± 6.5	44 ± 5.2
15 years (n = 37)		
$\dot{V}_{O_2}$ max, ml/min	3,328 ± 462	2,854 ± 359
$\dot{V}_{O_2}$ max (ml/kg · min)	54.4 ± 7.2	47 ± 6.2

Table 9. Anaerobic threshold of subjects who practice team games and rhythmic events (mean ± SD)

Group	Team games	Rhythmic events
12 years (n = 38)		
Percentage of $\dot{V}_{O_2}$ max	54 ± 11	54 ± 8
Work rate, W	64 ± 29	50 ± 8
13 years (n = 54)		
Percentage of $\dot{V}_{O_2}$ max	54 ± 6	51 ± 4
Work rate, W	80 ± 18	73 ± 11
14 years (n = 53)		
Percentage of $\dot{V}_{O_2}$ max	55 ± 7	54 ± 4
Work rate, W	85 ± 13	86 ± 12
15 years (n = 37)		
percentage of $\dot{V}_{O_2}$ max	57 ± 4	52 ± 2
Work rate W	92 ± 12	89.5 ± 20

Table 10. Maximal oxygen intake, maximal oxygen debt and anaerobic threshold of subjects from the 12- to 13-year age group at the experimental school and from sports schools (mean ± SD)

Group	$\dot{V}_{O_2}$ max ml/min	$\dot{V}_{O_2}$ max ml/kg · min	O_2 debt max		AT, % of $\dot{V}_{O_2}$ max
			ml	ml/kg	
Experimental school					
Measure I (n = 32)	2,205 ± 326	48.8 ± 7.9	4,995 ± 1,005	100 ± 19	58 ± 5.6
Measure II (n = 32)	2,813 ± 360	58.2 ± 3.9	6,145 ± 775	126 ± 8.6	68 ± 6
Track and field (n = 15)	2,246 ± 517	48.5 ± 7.2	5,500 ± 697	120 ± 7.8	50 ± 5.6
Basketball (n = 19)	2,800 ± 398	45.8 ± 5.6	5,025 ± 576	82 ± 28	54 ± 6
Baseball (n = 21)	2,715 ± 452	50.3 ± 5.6	5,405 ± 876	97 ± 31	48 ± 6
Boxing (n = 20)	2,617 ± 350	54 ± 6	6,700 ± 820	142.5 ± 20	60 ± 6

subjects who were assessed twice were grouped into two age categories (12–13 years and 14–15 years) (tables 10, 11).

The physical activities at the *experimental center* apparently have more impact upon aerobic power than the other motor regimens investigated. In terms of the $\dot{V}_{O_2}$ max/kg, only boxers excel over the special school students, by 4 ml/kg · min ($p < 0.05$). It seems that over this age range, *close specialization* has excessive unilateral influences in the development of the child's body. Students at the *special center* achieve a 20.3% increase of $\dot{V}_{O_2}$ max/kg, a very large increase for only 9 months of training. The increase of aerobic power in the age group 14–15 years is similar, 19.7%. The data for anaerobic capacity suggest that the motor regimen of the *special center* leads mainly to the development of cardio-respiratory power. We come to this conclusion since the O_2 debt (max/kg) increases by 26% in the 12- to 13-year age group, but in the 14- to 15-year age group the gain is only 3%. These peculiarities in the impact of different motor regimens upon the developing body at specific ages will find application to the practice of physical education and sports for children and adolescents. The child is not a miniature adult, but an organism subjected to the laws of biological development. It differs from the adult in both morphological and physiological characteristics, evolving irregularly and with periodic 'spurts'. If physical

Table 11. Maximal oxygen intake, maximal oxygen debt and Anaerobic threshold of subjects from the 14- to 15-year age group at the experimental school and from the sports school (mean ± SD)

Group	$\dot{V}_{O_2}$ max ml/min	$\dot{V}_{O_2}$ max ml/kg · min	O_2 debt max		AT, % of $\dot{V}_{O_2}$ max
			ml	ml/kg	
Experimental school					
Measure I (n = 18)	2,166 ± 202	44 ± 6.8	4,955 ± 952	96 ± 18	55 ± 5.9
Measure II (n = 18)	2,705 ± 220	52.7 ± 5	5,025 ± 1,324	99 ± 28	66 ± 8.2
Sports school					
Basketball (n = 8)	2,898 ± 331	49 ± 6	5,690 ± 1,806	92 ± 28	50 ± 5.8
Baseball (n = 20)	3,015 ± 474	48.5 ± 6.8	6,044 ± 1,906	96 ± 30	55 ± 5.6

education and sports are to favor such development, the proposed motor regimens should be assessed according to the effects that they cause. Our findings suggest that the practice of only one sport at the peak period of puberty is not the ideal method of developing aerobic power. The acceleration of physical development seen in recent generations (particularly body mass) encourages a substantial intensification of the physical exercise practiced since an early age. On the other hand, young people are entering the international arena at progressively lower ages, apparently without adverse effects, implying that a physical overload can be tolerated from a very early age.

Further study of different motor regimens is needed in order to determine what programme is the most suitable for each age group. Our investigations indicate that the *schools in the countryside* favor cardiovascular development relative to students from schools in urban areas. Our findings also confirm the favorable influence of sports practice on the child's body. Nevertheless, a varied motor regimen of *team games* seems to agree best with the needs of the body at these ages. In this context, the multilateral regimen of the *experimental center*, with *close specialization*, seems to have an optimal effect on morphological and physiological development.

References

1 Serralta MF: Estudio ergonómico de una azada apropiada a las caracteristicas de los estudiantes de las secundarias básicas en el campo. Universidad de La Habana, Facultad de Tecnología, 1975, pp 68–80.

2   Wahlund H: Determination of physical working capacity. Acta Med Scand 1948;132:1–78.
3   Yañez J, Guminsky A: Un métoda analitico para la determinación de las fracciones alactácida y lactácida de la deuda de oxígeno. Bol Científico-técn, INDER, Cuba 1978;2:19–24.
4   Brooks GA: 'Anaerobic threshold': Review of the concept and directions for future research. Med Sci Sports Exerc 1985;17:22–31.
5   Tanner JM: Growth at Adolescence. Oxford, Blackwell, 1955.
6   Shephard RJ: The working capacity of school children; in: Frontiers of Fitness. Springfield, Thomas, 1971, pp 319–344.
7   Macek M, Vavra J: Prolonged exercise in children. Acta Pediatr Belg 1974;28:13–18.
8   Raczek J, Brehmer R: Znaczenie okreslania progow przemian tlenowych i beztlenowych dla sterowania treiningien wytrzymalosciowyn. Sport Wyczynowy 1980;4:3–14.

Nelson Arbesú, PhD, Institute of Nutrition, Laboratory of Physiology, Infanta 1158, Havana 10300 (Cuba)

Anthropometric, Functional and Psychological Characteristics of Eight-Year-Old Brazilian Children from Low Socioeconomic Status

M.B. Rocha Ferreira[a], *Robert M. Malina*[b], *Lydia L. Rocha*[c]

[a]Faculty of Physical Education, UNICAMP, São Paulo, Brazil; [b]Department of Kinesiology and Health Education, University of Texas, Austin, Tex., USA; [c]Institute of Psychology, USP, São Paulo, Brazil

Introduction

Protein-energy malnutrition (PEM) is an important public health problem. Depending on the intensity, duration and timing of this stress, its effects range from mild to severe [Malina, 1984]. The immediate cause is nutritional, but there are important contributing factors, social, cultural, economic, and perhaps political. Infectious and parasitic diseases are additional related factors. Nevertheless, the primary problems are deficiencies of energy and protein [Bengoa, 1971; Marcondes et al., 1972; Cassidy, 1982].

PEM may give rise to a severe, and, occasionally, an irreversible impairment of health. It usually has mild to moderate effects on health, particularly when the stress is imposed in the early years of life; however, such changes are difficult to identify relative to clinical manifestations of the more severe forms of malnutrition. Characteristic features are stunted growth, delayed maturation, reduced muscle mass, and decreased physical working capacity [Bengoa, 1971; Himes, 1978; Buschang, 1980; Spurr et al., 1982, 1983; Spurr, 1983; Malina, 1984, 1985, 1986; Malina and Buschang, 1985; Johnston et al., 1985].

Malnutrition can also affect motor performance. Malnourished children generally show low scores in tests of speed, jumping, strength, long-distance runs and throwing [Malina, 1981, 1984, 1986; Malina and Buschang, 1985].

Many authors have suggested that severe undernutrition during infancy might have its greatest effect on growth of the brain. Retardation of intellectual development is therefore a likely consequence [Stoch and Smythe, 1963; Cabak and Najdanvic, 1965; Cravioto et al., 1966; Edwards and Craddock, 1973; Freeman et al., 1977]. In addition, chronic PEM can affect emotional functioning [Barrett and Radke-Yarrow, 1982]. Although nutrition may indeed be a factor in delayed intellectual development, it is difficult to isolate from other social and cultural factors.

The present chapter considers the interrelationships between growth status, physical performance and psychological characteristics in 145 eight-year-old Brazilian schoolchildren of both sexes, living in conditions of chronic mild to moderate undernutrition. Estimates of food intake have been made in an attempt to evaluate the nutritional status of these children.

Material and Methods

The children studied (61 girls and 84 boys) were attending a public school in Sao Paulo. They were all of low socioeconomic status (SES), and in general appeared to live in conditions of mild to moderate undernutrition. Nevertheless, the majority of the families earned from three to four times the minimum monthly salary in Brazil (equivalent to US $ 36 in 1985).

The anthropometric dimensions measured included body mass, height, selected circumferences (arm and calf) and six skinfolds (biceps, triceps, subscapular, suprailiac, abdominal and medial calf). The physical performance tests comprised a standing long jump, a shuttle run, a 50-meter dash, a 9-minute run, and a measure of maximal handgrip force made by a Stoelting dynamometer with adjustable size of grip. The psychological development was assessed in terms of an intelligence quotient (Terman-Merrill Stanford-Binet scale, Form M [Terman and Merrill, 1966]), visuomotor perception (Bender's graphic test of perceptual organization, as revised by Santucci and Galifret-Granjon [1963]), and body image (Goodenough's 'Draw a Man' test).

The food intake was determined at home and at school and socioeconomic information was collected by interviewing the adult responsible for each child [Rocha-Ferreira, 1987]. The recall method estimated food intake over a 24-hour period, and Brazilian tables elaborated by Tudisco et al. [1978] were used to transform the food intake into grams. The tables of FIBGE [1977] converted this information to measures of energy and protein intake.

Results

The means and standard deviations of the anthropometric, performance and psychological data are summarized in table 1. There were no significant differences in size or circumferences between the boys and the girls, but the girls had thicker skinfolds. The boys performed better than the girls on all tests of performance and the measure of intelligence; they also had a better visual-motor perception and body image than the girls.

Table 2 presents sex-specific zero-order correlations between measures of growth, physical performance and psychological characteristics. Stature and body mass were significantly correlated with other anthropometric data in both sexes. Correlations among the motor performance scores

Table 1. Means, SD and t tests (2-tail probability) for anthropometric, performance and psychological characteristics by sex

Variables	Girls		Boys		t
	mean	SD	mean	SD	
Age, years	8.5	0.3	8.5	0.3	−0.4
Weight, kg	25.9	4.9	25.5	0.6	0.6
Height, cm	125.8	5.9	126.2	5.5	−0.4
Circumferences, cm					
Arm relaxed	17.8	2.0	17.4	1.0	1.3
Arm tensed	19.4	1.9	19.2	1.1	0.6
Calf	25.2	2.3	24.7	1.5	1.7
Estimated mid-arm muscle	14.6	1.1	14.9	1.0	−1.4
Skinfolds, mm					
Triceps	10.0	4.0	8.1	2.0	3.6**
Biceps	5.0	2.0	4.0	0.9	4.1**
Abdominal	6.7	4.0	5.2	1.7	3.0**
Suprailiac	5.5	4.3	4.1	1.2	2.8**
Subscapular	6.5	3.9	5.0	1.4	3.4**
Calf	9.1	3.7	7.0	2.1	4.4**
Performance					
Grip, kg	12.2	2.3	13.2	2.3	−2.9**
Jump, cm	116.3	12.1	128.8	15.7	−5.1**
Speed, s	11.1	0.9	10.6	1.0	3.7**
Agility, s	13.9	1.1	13.2	1.0	4.0**
Run, m	1,402.4	172.6	1,573.9	196.7	−5.2**
Psychological tests					
IQ	83.4	11.3	87.5	12.8	−2.0*
Perception	25.2	6.7	28.0	7.8	−2.2*
Mental age	7.1	1.0	7.5	1.1	−2.1*
Body image	3.3	1.0	3.0	0.8	−2.1*

* $p < 0.05$; ** $p < 0.01$.

ranged from low to moderate, although most were statistically significant. Correlations among the psychological test scores ranged from moderate to high.

Stature, body mass, estimated mid-arm muscle circumference and handgrip force were significantly but moderately correlated. Correlations between stature, body mass, mid-arm muscle circumference and the three running tests (speed, agility and endurance) tended to be positive, but were

Table 2. Zero-order correlations among anthropometric dimensions, physical performance, and psychological tests in 8-year-old lower socioeconomic girls (n = 60) and boys (n = 84)

	1	2	3	4	5	6	7	8	9	10	11	12	13	14	15
Girls															
1 Weight	–	0.69**	0.73**	0.89**	0.84**	0.88**	0.31**	−0.09	−0.03	0.13	−0.24*	0.10	0.17	0.18	0.05
2 Height		–	0.41**	0.41**	0.37**	0.43**	0.52**	0.12	0.14	0.31*	0.02	0.08	0.18	0.20	0.07
3 Arm muscle			–	0.56**	0.52**	0.57**	0.32**	0.10	0.12	0.26*	−0.08	0.09	0.12	0.15	0.05
4 Sksum				–	0.98**	0.97**	0.05	−0.18	−0.20	−0.08	−0.41**	0.11	0.12	0.20	0.06
5 Skextrem					–	0.90**	0.07	−0.18	−0.21	−0.07	−0.35**	0.12	0.11	0.13	0.08
6 Sktrunk						–	0.04	−0.18	−0.19	−0.09	−0.43**	0.10	0.12	0.24*	0.03
7 Grip							–	0.45**	0.46**	0.38**	0.16	0.35**	0.11	0.06	0.40**
8 Jump								–	0.46**	0.41**	0.32**	0.22*	0.05	0.02	0.23*
9 Speed									–	0.55**	0.43**	0.25*	0.15	0.15	0.28**
10 Agility										–	0.39**	0.08	0.10	−0.01	0.14
11 Run											–	0.07	0.10	−0.06	0.13
12 IQ												–	0.42**	0.47**	0.96**
13 Perception													–	0.51**	0.39**
14 Body image														–	0.41**
15 Mental age															–
Boys															
1 Weight	–	0.82**	0.64**	0.40**	0.34**	0.40**	0.48**	−0.01	0.12	0.09	−0.06	−0.06	0.06	0.16	0.06
2 Height		–	0.49**	0.07	0.05	0.07	0.47**	0.03	0.21	0.20	0.11	0.00	−0.02	0.17	0.04
3 Arm muscle			–	−0.07	−0.15	0.05	0.51**	0.02	0.17	0.04	0.09	−0.11	−0.16	−0.12	−0.06
4 Sksum				–	0.93**	0.92**	−0.01	−0.24**	−0.36**	−0.19*	−0.43**	0.21**	0.24**	0.15	0.13
5 Skextrem					–	0.71**	−0.04	−0.23**	−0.30*	−0.14	−0.42**	0.29**	0.32**	0.19	0.22*
6 Sktrunk						–	0.02	−0.22*	−0.37**	−0.23**	−0.38**	0.10	0.11	0.08	0.01
7 Grip							–	0.29**	0.38**	0.19*	0.07	0.13	0.06	0.06	0.19*
8 Jump								–	0.57**	0.47*	0.27**	0.17	0.11	−0.02	0.19*
9 Speed									–	0.38*	0.49**	0.05	0.05	0.01	0.14
10 Agility										–	0.26*	0.20*	0.13	0.19*	0.26**
11 Run											–	−0.18	−0.16	−0.10	−0.13
12 IQ												–	0.59**	0.40**	0.96**
13 Perception													–	0.33**	0.59**
14 Body image														–	0.39**
15 Mental age															–

Arm muscle = Estimated midarm muscle circumference; Sksum = sum of six skinfold thicknesses; Skextrem = sum of skinfold extremities; Sktrunk = sum of skinfold trunk. Signs for the speed and agility tests are reversed.
* $p < 0.05$; ** $p < 0.01$.

low in both sexes. In contrast, correlations between the sum of skinfold thicknesses and the jump, dash, agility and endurance runs were all negative, but low to moderate in magnitude; skinfold thicknesses were unrelated to handgrip force.

Few anthropometric and physical performance measures were significantly related to psychological scores. Stature and body mass showed low correlations with IQ, perception, body image and mental age. The sum of skinfolds had a low positive correlation to all psychological scores in both sexes, although correlations were weaker in the girls than in the boys.

The correlations between psychological scores and physical performance were also weak; low to moderate positive correlations were seen between mental age and handgrip force, jump, speed, agility and running distance scores. IQ had a low positive correlation with physical performance. Visual-motor perception had a low to moderate positive correlation with agility and speed scores.

Second-order partial correlations controlling for age and stature or body mass are summarized in table 3. After controlling for stature and age, all correlations between body mass and motor performance were negative, with the exception of handgrip force in the boys. Although the coefficients were low to moderate, body mass had a negative effect on both jumping and running tasks. On the other hand, all second-order partial correlations with stature and motor performance were positive after controlling for age and body mass. Again, the coefficients were low to moderate, but they indicated that stature had a positive effect on motor performance.

Second-order partial correlations between psychological characteristics and body mass or stature both approached zero.

Estimates of the daily energy and protein intakes are shown in table 4. Boys tended to consume more energy and protein than girls, whether results were expressed per day (kcal, 1 kcal = 4.186 kJ) or per unit of body mass (kcal/kg/day). Both sexes met the recommended minimum protein intakes, but appeared to be deficient in total energy intake. Correlations between the estimated energy intake and anthropometric characteristics were weak but positive. However there was no clear pattern of association between protein intakes and measures of growth. Neither energy nor protein intake seemed related to physical or psychological performance.

Discussion

The children studied here on average had a lower height and body mass than other children from the Sao Paulo area [Marques et al., 1975; Sessa et al., 1978; Barbanti, 1982; Oliveira and Franca, 1983; Rocha-

Table 3. Second-order partial correlations between age and body size, and motor performance

Variables	Motor performance and weight, controlling for age and stature		Motor performance and stature, controlling for age and weight	
	girls	boys	girls	boys
Performance				
Grip	−0.03	0.19*	0.42**	0.14
Jump	−0.22	−0.06	0.24*	0.07
Speed	−0.13	−0.07	0.18	0.14
Agility	−0.07	−0.12	0.28*	0.19*
Run	−0.31*	−0.26*	0.24*	0.27*
Psychological tests				
IQ	0.05	0.10	0.02	−0.08
Perception	0.04	0.03	0.07	−0.07
Mental age	0.00	0.03	0.14	0.07
Body image	0.03	0.13	0.11	−0.11

Signs for the speed and agility tests are reversed since a lower time is a better performance.

Table 4. Estimated means ± SDs of energy and protein intakes of lower SES children and recommended dietary allowances

	Energy		Protein	
	kcal	kcal/kg	g	g/kg
Lower SES				
girls	1,286 ± 434.9	50.7 ± 17.9	49.3 ± 19.0	1.9 ± 0.8
boys	1,320 ± 428.3	52.3 ± 17.4	51.3 ± 16.6	2.0 ± 0.7
RDA[a]				
Girls	2,110	76.0	25	0.9
Boys	2,260	79.0	25	0.9

[a] WHO/FAO [1973].

Ferreira, 1987]. The only exception to this generalization is that the girls were slightly heavier and taller than the sample of Sessa et al. [1978]. The other children were drawn from low/middle SES, except that Rocha-Ferreira [1987] reported data for both lower and upper SES.

When the centiles for height and body mass were compared with US reference data [Hamill et al., 1977], the children from the present study were seen to lie between the 27th and 44th US centiles for height, and the 28th to 55th centiles for body mass. In general, differences from other Brazilian samples were quite small, suggesting that the present sample was not as undernourished as assumed. However, boys in the present study had the lowest centiles, suggesting that they may be more responsive to an adverse environment than are girls [Bielicki and Charzewski, 1977; Brauer, 1982; Malina and Buschang, 1985].

Body size and proportions showed low-to-moderate positive correlations with motor performance and handgrip force, as previously reported [Seils, 1951; Rarick and Oyster, 1964; Malina, 1975; Malina and Buschang, 1985]. The relationship was low to moderate, because other variables affect both dimensions and motor performance, for instance, prior physical activity, physical education programmes and nutritional status. Consistent with other studies, heavier children had a larger handgrip force after control of data for age and height [Malina, 1975; Malina and Buschang, 1985; Malina et al., 1987]. However, heavier children also had a poorer motor performance, again in keeping with the data of Malina [1975].

When age and body mass were controlled, stature was positively correlated with jumping, agility and 9-minute run scores; again, these findings are in keeping with the observations of Malina and Buschang [1985] and Malina et al. [1987] for students aged 6–14 years.

The tendency for low SES children to score poorly on psychological tests is well established [Cravioto et al., 1966; Silva Carmo, 1969; Edwards and Craddock, 1973; Freeman et al., 1977; Barrett and Radke-Yarrow, 1982; Oakland and Ramos-Cancel, 1985]. Boys also perform better than girls in psychological tests, a fact previously stressed in the review of Oakland and Ramos-Cancel [1985].

Height and body mass showed little relationship to psychological performance. Presumably, variables such as poor conditions for the development of language and reasoning, inferior housing, lack of medical care and other factors related to poverty had a larger impact on scores in the psychological tests.

The immediate energy and protein intakes showed little relationship to growth status, physical performance or psychological performance. This is perhaps not surprising, given the limited accuracy of the estimates, and the fact that prior nutrient intake is probably a more significant variable.

In summary, the growth, performance and psychological status of 8-year-old lower SES children are consistent with the marginal economic circumstances of their families. Their growth status and motor performance compares favorably with other low SES Brazilian children, but is inferior to that of upper SES children, except in the endurance run. The overall poor performance on the psychological tests probably reflects the limited intellectual stimulation received in their home environment.

References

Barbanti, V.J.: Comparative study of select anthropometric and physical fitness measurements of Brazilian and American School children; doctoral diss., University of Iowa, Iowa city (1982).
Barrett, D.E.; Radke-Yarrow M.: Chronic malnutrition and child behavior: effects of early caloric supplementation on social and emotional functioning at school age. Devl. Psychol. *18:* 541–556 (1982).
Bengoa, J.M.: Recent trends in the public health aspects of protein-calorie malnutrition. Cajanus *4:* 141–159 (1971).
Bielicki, T.; Charzewski, J.: Sex differences in the magnitude of statural gains of offspring over parents. Hum. Biol. *49:* 265–277 (1977).
Brauer, G.W.: Size sexual dimorphism and secular trend: Indicators of subclinical malnutrition? In Hall, Sexual dimorphism in *Homo sapiens*, pp. 245–259 (Praeger, New York 1982).
Buschang, P.H.: Growth status and rate of school children 6 to 13 years of age in a rural Zapotec-speaking community in the Valley of Oaxaca, Mexico; doctoral diss., University of Texas at Austin (1980).
Cebak, V.; Najdanvic, R.: Effect of undernutrition in early life on physical and mental development. Archs Dis. Childh. *40:* 532–539 (1965).
Cassidy, C.M.: Protein-energy malnutrition as a culture-bound syndrome. Culture Med. Psychiat. *6:* 325–345 (1982).
Cravioto, J.; De Licardie, E.R.; Birch, H.G.: Nutrition, growth, and neurointegrative development: An experimental and ecologic study. Pediatrics *38:* 319–372 (1966).
Edwards, L.D.; Craddock, L.J.: Malnutrition and intellectual development: A study in school age aboriginal children at Walgett, N.S.W. Med. J. Aust. *1:* 880–884 (1973).
Freeman, H.E.; Klein, R.E.; Kagan, J.; Yarbrough, C.: Relations between nutrition and cognition in rural Guatemala. Am. J. Publ. Hlth *67:* 233–239 (1977).
Fundaçao Instituto Brasileiro de Geografia e Estatistica – FIBGE' Estudo Nacional da Despesa Familiar – ENDEF – Tabelas de Composiçao dos Alimentos. Secretaria de Planejamento da Presidencia da Republica, Rio de Janeiro (1977).
Hamill, P.V.V.; Johnson, C.L.; Roche, A.F.: NCHS growth curves for children, birth–18 years, United States. Vital and Health Statistics Series 11, No. 165 (United States Government Printing Office, Washington 1977).
Himes, J.H.: Bone growth and development in protein-calorie malnutrition. Wld Rev. Nutr. Dietet. *28:* 143–187 (1978).
Johnston, F.E.; Low, S.M.; Baessa, Y.; MacVean, R.B.: Growth status of disadvantaged urban Guatemalan children of a resettled community. Am. J. phys. Anthropol. *68:* 215–224 (1985).

Malina, R.M.: Anthropometric correlates of strength and motor performance. Exerc. Sport Sci. Rev. *3:* 249–274 (1975).
Malina, R.M.: Growth and performance of Latin American children. Prepared for the Kinanthropometry Section of the PanAmerican Congress of Sports Medicine and Exercise, Miami, 1981.
Malina, R.M.: Physical activity and motor development/performance in populations nutritionally at risk; in Pollitt, Amante, Energy intake and activity, pp. 285–302 (Liss, New York 1984).
Malina, R.M.: Growth and physical performance of Latin American children and youth: Socioeconomic and nutritional contrasts. Coll. Anthropol. *9:* 9–31 (1985).
Malina, R.M.: Motor development and performance of children and youth in undernourished populations; in Katch, Sport Health and Nutrition, pp. 213–226 (Human Kinetics Publishers, Champaign 1986).
Malina, R.M.; Buschang, P.H.: Growth, strength and motor performance of Zapotec children, Oaxaca, Mexico. Hum. Biol. *57:* 163–181 (1985).
Malina, R.M.; Little, B.B.; Shoup, R.F.; Buschang, P.H.: Adaptive significance of small body size: strength and motor performance of school children in Mexico and Papua New Guinea. Am. J. phys. Anthropol. *73:* 489–499 (1987).
Marcondes, E.; Monteiro, D.M.; Barbieri, D.; Quarentei, G.; Yunes, J.; Campos, J.V.M.; Setian, N.: Desnutriçao (Sarvier, Sao Paulo 1972).
Marques, R.M.; Berquo, E.; Hegg, R.; Colli, A.S.; Zacchi, M.A.S.: Crescimento de ninos Brasilenos: peso y altura en relacion con la edade y el sexo y la influencia de factores socioeconomicos. Organizacion Panamericana de la Salud, Organizacion Mundial de la Salud, Publ. Cientifica, No. 309 (1975).
Oakland, T.D.; Ramos-Cancel, M.: Educational perspectives on Hispanic children from Hispanic journals: A view from Latin America. J. multiling. multicult. Dev. *6:* 67–82 (1985).
Oliveira, J.A.; França, N.M.: Diametros e circunferencias de escolares de 7 a 11 anos. Ill Congresso Brasileiro de Ciencias do Esporte, Guarulhos, 1983, p. 139.
Rarick, G.L.; Oyster, N.: Physical maturity, muscular strength, and motor performance of young school-age boys. Res. Q. *35:* 523–531 (1964).
Rocha Ferreira, M.B.: Estado Nutricional e aptidao fisica em pre-escolares. Tese de mestrado apresentada na Escola de Educaçao Fisica da Universidade de Sao Paulo, 1979.
Rocha Ferreira, M.B.: Growth, physical performance and psychological characteristics of eight year old Brazilian school children from low socioeconomic background; doctoral diss., University of Texas at Austin, 1987.
Santucci, H.; Galifret-Granjon, N.: Prova grafica de organizacion perceptiva; in Zazzo, Manual para el examen psicologico del nino, pp. 177–208 (Editorial Kapelusz, Buenos Aires 1963).
Seils, L.G.: The relationship between measures of physical growth and gross motor performance of primary-grade school children. Res. Q. *22:* 244–260 (1951).
Sessa, M.; Matsudo, V.K.R.; Vivolo, M.A.; Tarapanoff, A.M.: Desenvolvimento da força de membros inferiores em escolares 7 a 18 anos em funçao de sexo, idade, peso, altura e atividade fisica, Anais Vl Simp. Ciencias do Esporte, Sao Caetano do Sul, Sao Paulo, 1978, pp. 31–35.
Silva Carmo, H.M.: O problema dos repetentes da la serie primaria nos grupos escolares de Sao Paulo, Report, Departamento de Psicologia, Universidade de Sao Paulo, 1986.
Spurr, G.B.: Nutritional status and physical work capacity. Yearb. phys. Anthropol. *26:* 1–34 (1983).

Spurr, G.B.; Reina, J.C.; Barac-Nieto, M.; Maksud, M.: Maximum oxygen consumption of nutritionally normal white, mestizo and black Colombian boys 6–16 years of age. Hum. Biol. *54:* 553–557 (1982).

Spurr, G.B.; Reina, J.C.; Barac-Nieto, M.: Marginal malnutrition in school-aged Colombian boys: anthropometry and maturation. Am. J. clin. Nutr. *37:* 119–132 (1983).

Stoch, M.B.; Smythe, P.M.: Does undernutrition during infancy inhibit brain growth and subsequent intellectual development? Archs Dis. Childh. *38:* 546–552 (1963).

Terman, L.M.; Merrill, M.A.: Medida de la inteligencia, metodo para el empleo de las pruevas de Stanford-Binet nuevamente revisadas (Espasa-Calpe, Madrid 1966).

Tudisco, E.S.; Manoel, N.J.; Sigulem, D.M.: Guia para avaliaçao da dieta do paciente em consulta ambulatorial, Resumos do Congr. Int. Nutriçao, Rio de Janeiro, 1978, p. 272.

Zazzo, R.: Manual para el examen psiologico del nino (Kapelusz, Buenos Aires 1963).

M.B. Rocha Ferreira, PhD, Faculdade de Educação Física, UNICAMP, Campinas, São Paulo 13081 (Brazil)

Estimated Body Composition and Strength of Chronically Mild-to-Moderately Undernourished Rural Boys in Southern Mexico[1]

Robert M. Malina[a] *Bertis B. Little*[b], *Peter H. Buschang*[c]

[a]Department of Kinesiology and Health Education, University of Texas, Austin, Tex., USA; [b]Division of Clinical Genetics, Department of Obstetrics and Gynecology, University of Texas, Southwestern Medical Center, and [c]Baylor College of Dentistry, Department of Orthodontics, Dallas, Tex., USA

Introduction

Changes in body composition with severe malnutrition are reasonably well documented in infancy and early childhood [1, 2]. On the other hand, detailed studies of the body composition of children living under conditions of chronic, mild-to-moderate undernutrition are not available, although the anthropometric characteristics of these children, particularly stature, body mass, arm and estimated midarm muscle circumferences, and skinfolds, are well documented [3–8]. School-age children in communities with chronic, mild-to-moderate undernutrition represent, to a large extent, the survivors of an early childhood that was characterized by nutritional and infectious disease stress. More importantly, perhaps, they will be the next generation of economically active adults, and the long-term consequences of chronic undernutrition experienced in early childhood may result in depressed physical work capacity in adulthood [9, 10]

While there is some information on the body composition of adult males suffering from chronic undernutrition relative to physical work capacity [11, 12], corresponding data for school-age children are lacking. Chronic, mild-to-moderate undernutrition is associated with reduced strength and motor performance; however, performance levels are to some extent commensurate with reduced body size [13, 14]. Data on the relationship between performance and body composition of undernourished chil-

---

[1] This research was supported in part by the Institute of Latin American Studies of the University of Texas at Austin. Part of the data was collected under a grant from the National Science Foundation, BNS 78-10642.

dren are lacking. In a sample of 126 boys 9 through 15 years of age living under conditions of chronic undernutrition, relationships between estimated absolute body composition and motor performance (run, jump, throw) and grip strength were generally similar to those for well-nourished boys. However, relationships between relative body composition and performance differed; relative fatness had a negligible effect on the performance of undernourished boys [15]. This observation suggested that greater relative fatness in undernourished individuals may be indicative of a relatively better nutritional status. This chapter extends our earlier analysis of body composition and performance in boys living under chronically marginal nutritional circumstances. It specifically considers relationships between estimated body composition and muscular strength in a larger sample of boys 9 through 14 years of age. The growth, body composition and strength of the Mexican boys are compared to a better nourished sample of Mexican American boys living in Texas.

The Community

In 1968, 1978 and 1979 we surveyed the growth status of schoolchildren in a rural, Zapotec-speaking community in the Valley of Oaxaca in southern Mexico [16–19]. The community is located in the Valley of Oaxaca, northwest of the city of Oaxaca de Juarez, the capital of the State of Oaxaca. It is a subsistence agricultural community, comprising about 1,250 residents in 1965 and about 1,700 residents in 1978. Life-style in the community is typical of traditional Middle-American communities, that is, largely endogamous, with farmers working small plots of individually owned land. There is no doctor in the community, although a midwife with some public health experience moved into the community in the late 1970s.

The nutritional status of the community is at best marginal. Estimated energy, protein, vitamin and mineral intakes are considerably lower than recommendations for Mexico [20]. The majority of households consume only 5 food items: *tortillas* (corn-based bread), *cafe negro* (black coffee), *salsa* (sauce comprised of tomatoes, peppers and spices), *frijoles* (beans), and *pan* (flour-based bread) on a more or less regular basis [21]. The marginal nutritional status is reflected in high crude death rates and an especially high infant mortality rate, estimated at about 150/1,000 live births in the 1970s. This rate is about three times that for all of Mexico. The marginal nutritional conditions are also reflected in the statures and body masses of schoolchildren which only approximate the 5th percentiles of reference data for well-nourished children [22]. Further, there is no indication for either short-term (1968 to 1978) or long-term (since the turn of the century) secular change in stature in the community [17, 23]. These conditions are representative of the relatively poor nutritional status which characterizes the rural south of Mexico, including the state of Oaxaca.

Methods

This study considers Zapotec boys 9 through 14 years of age (because this is the age range of the sample upon which our equations for the prediction of body composition are derived, see below). During surveys in the community in 1968, 1978 and 1979, a variety of anthropometric dimensions were taken on schoolchildren, including stature, body mass, four skinfolds (triceps, subscapular, midaxillary and suprailiac, Lange skinfold caliper), and right and left grip strength (Stoelting adjustable dynamometer). Since there were no significant differences in any of the measurements taken on the children in 1968, 1978 and 1979, the data were combined into a single mixed-longitudinal sample of 320 boys. The average of right- and left-hand grip strength was used as an indicator of muscular strength.

The four skinfolds were used to predict body density, which was subsequently converted to relative fatness and in turn to fat mass (FM) and fat-free mass (FFM). A regression equation for the prediction of body density was developed on a sample of 95 lower socioeconomic status Mexican American boys 9 through 14 years of age resident in Austin, Tex. [24]. Measurement procedures were the same in both studies, although different skinfold calipers were used in each study. Technical errors of measurement are shown in table 1.

The regression equation developed to predict body density utilized age and four skinfolds:

Density $(g/cm^3) = 1.0882 - 0.0012$ (triceps) $- 0.0005$ (suprailiac) $- 0.0011$ (midaxillary) $- 0.0003$ (age) $+ 0.0004$ (subscapular), where $R = 0.69$, $R^2 = 0.47$, and $SEE = 0.0092$. The standard error of estimate is within the range for studies of preadolescent and adolescent boys, $0.0075-0.0100$ g/cm^3 [25].

The predicted densities of the Zapotec boys and the measured densities of the Mexican American boys were converted to relative fatness, using the procedures of Lohman [26], which allow for the changing chemical composition of the FFM during growth. This involves the use of estimated densities of the FFM for different ages in deriving the formula to convert density to relative fatness. Relative fatness was subsequently used to estimate FM and FFM for each boy.

Results

Age-specific sample sizes, means and standard deviations for stature, body mass, the four skinfolds and grip strength of Zapotec boys are summarized in table 2. Compared to the Mexican American sample, the Zapotec boys are significantly shorter and lighter, with significantly thinner skinfolds and lower grip strength (fig. 1–3). Relative to reference data recommended by the World Health Organization (US reference data), the Zapotec boys are below the 5th percentiles for stature and body mass, while the Mexican American boys are at the 25th percentile for body mass and the 10th percentile for stature.

Table 1. Technical errors of measurement[1] in the studies of Zapotec and Mexican American boys

Measurement	Zapotec		Mexican American intraobserver
	intraobserver	interobserver	
Body mass, kg	0.10	0.20	0.48
Stature, cm	0.33	0.67	0.54
Skinfolds, mm			
Triceps	0.23	0.45	0.51
Subscapular	0.33	0.42	0.55
Midaxillary	0.27	0.42	0.43
Suprailiac	0.33	0.37	0.95

[1] The technical error of measurement ($\sigma_e = \sqrt{\Sigma d^2/2n}$): the square root of the sum of the squared differences of replicate measurements divided by twice the number of pairs [34]. The Mexican American data are from Zavaleta and Malina [24].

Table 2. Sample sizes, body size, grip strength and skinfold thicknesses of Zapotec boys (mean values ± SD)

Age group years	n	Stature cm	Body mass kg	Grip strength[1] kg
9+	61	121.1 ± 4.4	23.1 ± 2.2	11.7 ± 1.8
10+	64	125.1 ± 5.3	24.9 ± 3.2	13.5 ± 3.2
11+	58	129.4 ± 4.9	27.8 ± 3.7	15.8 ± 2.9
12+	53	133.2 ± 5.3	29.9 ± 3.2	17.5 ± 2.9
13+	49	136.1 ± 5.5	31.6 ± 4.1	18.0 ± 4.1
14+	35	144.0 ± 6.6	36.9 ± 4.9	21.3 ± 4.3

Age group years	Skinfold thicknesses, mm				
	triceps	subscapular	midaxillary	suprailiac	sum of 4
9+	6.0 ± 1.4	4.7 ± 0.8	3.7 ± 0.7	5.5 ± 1.9	20.0 ± 4.1
10+	6.3 ± 1.7	5.1 ± 1.1	3.8 ± 0.8	5.8 ± 1.9	21.0 ± 5.1
11+	6.2 ± 1.5	5.2 ± 1.1	3.9 ± 0.6	6.0 ± 1.8	21.3 ± 4.4
12+	6.5 ± 1.6	5.4 ± 1.0	4.1 ± 1.1	7.5 ± 3.5	23.5 ± 6.5
13+	6.0 ± 1.3	5.3 ± 0.8	3.9 ± 0.7	6.3 ± 1.5	21.6 ± 3.6
14+	6.0 ± 1.4	5.9 ± 0.9	4.2 ± 0.9	7.5 ± 2.1	23.6 ± 4.3

[1] Average of right and left grip strength

Body Composition and Strength

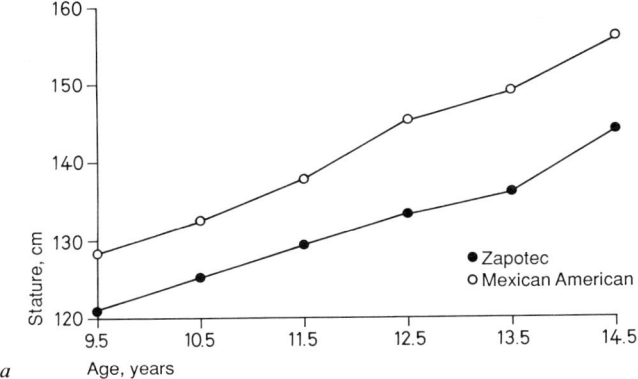

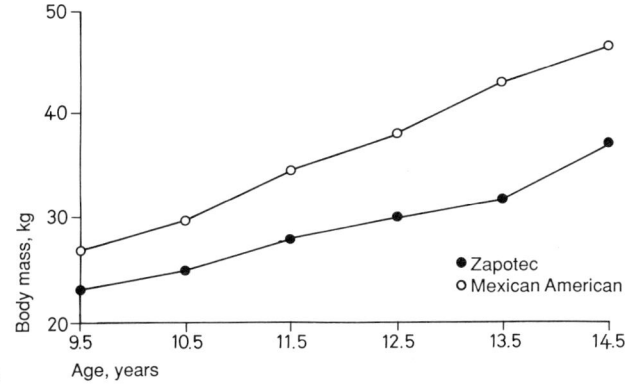

Fig. 1. Mean stature (*a*) and body mass (*b*) of Zapotec and Mexican American boys. All differences are significant ($p < 0.001$).

The predicted body composition of rural Zapotec boys is shown in table 3, while comparisons with Mexican American boys are shown in figure 4. The Zapotec boys have a smaller FFM and FM reflecting their smaller body size. The differences between Zapotec and Mexican American boys are significant at all ages except 9 years for FM. Zapotec boys also have less relative fatness, but the differences are not significant at 9 and 12 years of age.

Strength per unit body mass and per unit FFM is indicated in figure 5. The reduced strength of Zapotec boys is essentially a function of their reduced body mass. Grip strength per unit of body mass does not differ significantly between Zapotec and Mexican American boys from 10

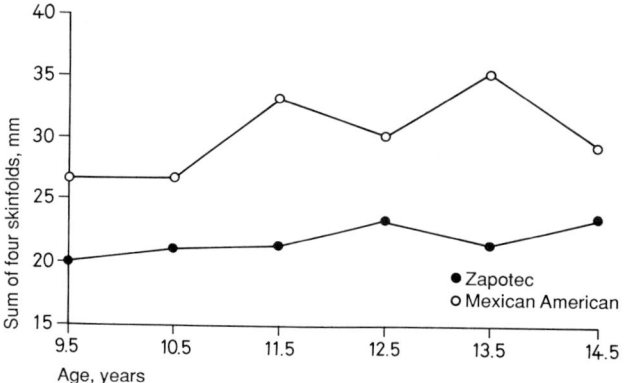

Fig. 2. Mean sum of four skinfolds (triceps, subscapular, midaxillary and medial calf) in Zapotec and Mexican American boys. All differences are significant ($p < 0.05$ 14 years, $p < 0.01$ 12 years, $p < 0.001$ all other ages).

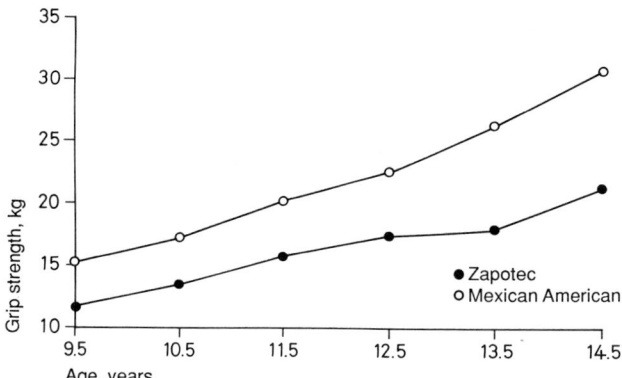

Fig. 3. Mean grip strength (average of right + left) of Zapotec and Mexican American boys. All differences are significant ($p < 0.001$).

through 13 years of age (fig. 5a), but per unit FFM, Zapotec boys have significantly less strength per unit FFM at all ages except 12 years (fig. 5b). The reduced strength per unit FFM in Zapotec boys suggests changes in the composition of the FFM and specifically in muscle mass associated with chronic protein-energy undernutrition.

Relationships among age, body size, body composition and grip strength are summarized in table 4. Correlations between grip strength and body mass, stature and FFM are high ($r > 0.80$), but are generally lower in

Table 3. Estimated body composition of Zapotec boys[1] (mean values ± SD)

Age group years	Body density g/cm³	Percent fat %	Fat mass kg	Fat free mass kg
9+	1.0734 ± 0.0026	4.8 ± 1.2	1.1 ± 0.3	22.0 ± 2.0
10+	1.0725 ± 0.0032	5.2 ± 1.5	1.3 ± 0.5	23.5 ± 2.9
11+	1.0723 ± 0.0026	6.6 ± 1.2	1.9 ± 0.5	25.9 ± 3.3
12+	1.0707 ± 0.0040	7.4 ± 1.8	2.2 ± 0.7	27.6 ± 2.9
13+	1.0717 ± 0.0025	9.6 ± 1.1	3.1 ± 0.6	28.6 ± 3.6
14+	1.0707 ± 0.0027	10.1 ± 1.2	3.7 ± 0.7	33.1 ± 4.3

[1] Sample sizes are as in table 2.

the undernourished Zapotec boys than in the better nourished Mexican American boys. In contrast, correlations between fatness and strength are considerably higher in Zapotec than in Mexican American boys.

Since body composition is related to age, body mass, and stature during growth, partial correlations between estimates of body composition and grip strength are shown in table 5. After controlling for age, body mass and stature, the correlation between FFM and grip strength is reduced, more so in Zapotec (from 0.84 to 0.27) than in Mexican American (from 0.93 to 0.59) boys. Absolute and relative fatness have a negative relationship with strength after age, body mass and stature are controlled, the negative effect being stronger in Mexican American (-0.51 and -0.59) than in Zapotec (-0.25 and -0.27) boys.

Discussion

The body composition estimates for chronically, mild-to-moderately undernourished Zapotec boys are based on body density predicted from skinfold thicknesses. The predicted densities were converted to relative fatness and in turn to FFM and FM using age- and sex-specific estimates of the density of the FFM during childhood to account for the chemical immaturity of children [26]. The constants, however, are from studies of well-nourished American children. Hence, one can inquire about the effects of chronic, mild-to-moderate protein-energy undernutrition on the composition of the FFM. Most of the available information is derived from patients with severe protein-energy malnutrition, who show reduced muscle mass, fibre size and energy metabolism [27], and changes in the hydration of fat-free tissue and in amount and distribution of proteins [1, 2]. Since

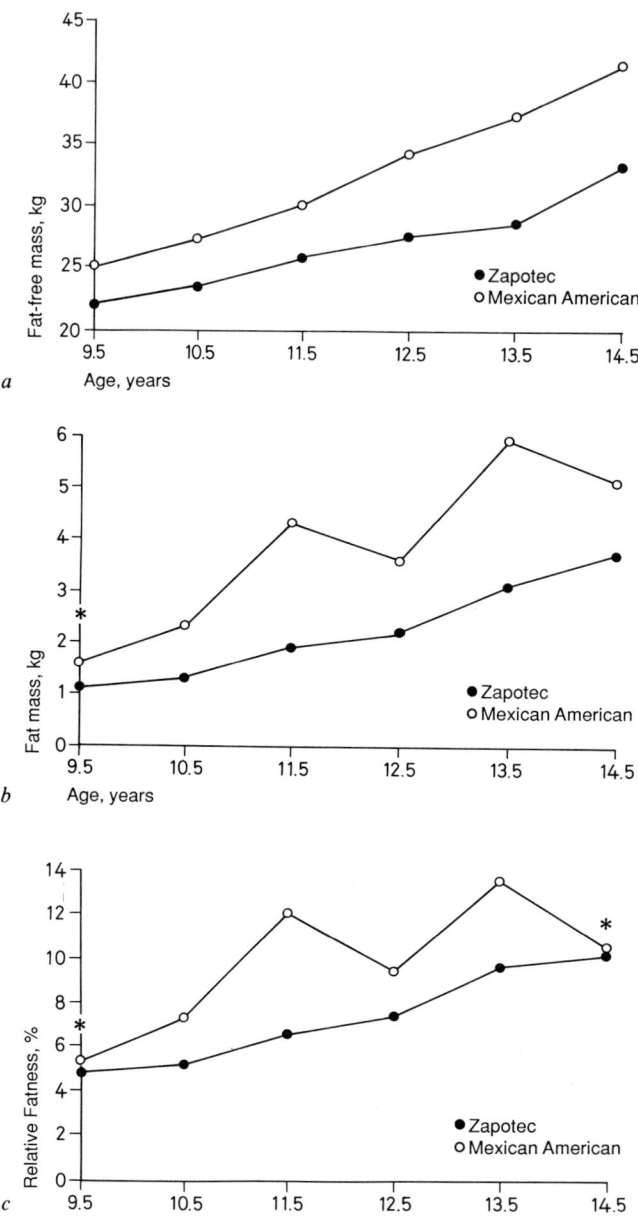

Fig. 4. Mean estimated body composition of Zapotec and Mexican American boys: FFM (*a*), FM (*b*), % fat (*c*). Differences indicated with an asterisk are not significant. All others are significant (p < 0.001 except FM 14 years and % fat 12 years, p < 0.05, and % fat 10 years, p < 0.01).

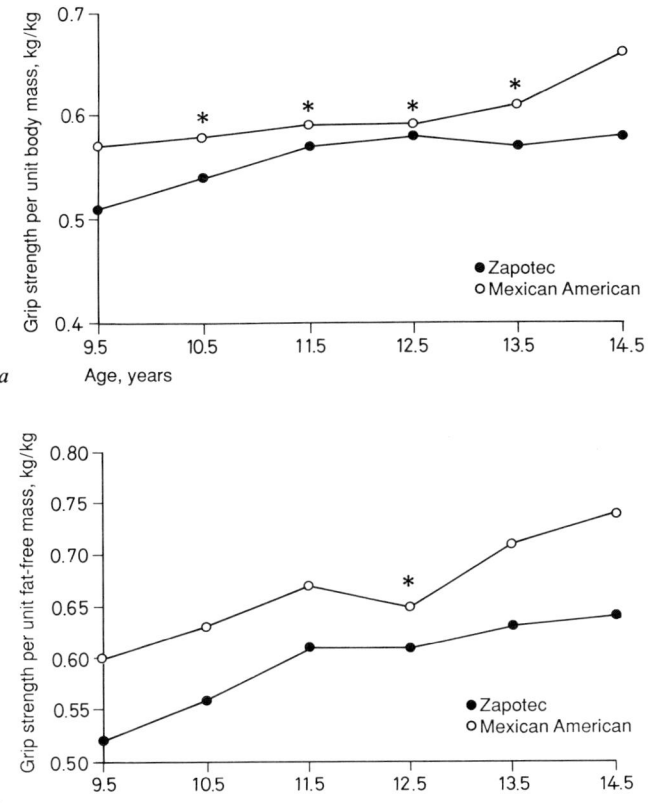

Fig. 5. Mean grip strength per unit body mass (*a*) and per unit FFM (*b*) in Zapotec and Mexican American boys. Differences indicated with an asterisk are not significant. All others are significant ($p < 0.01$ or $p < 0.05$).

protein-energy malnutrition represents a continuum from mild to severe in many populations in the developing world, it may be reasonable to assume that milder forms are associated with less severe changes in fat-free tissue. Reduced bone density may be an additional factor. Mild-to-moderate undernutrition is associated with reduced bone density and metacarpal cortical thickness [28]. Although metacarpal dimensions are related to body size, preschool Guatemalan children who suffered from mild-to-moderate protein-energy malnutrition [29] and school age Oaxaca children resident in an urban colonia with marginal health and nutritional circumstances [30] do in fact have smaller metacarpal dimensions than would be expected for

Table 4. Correlations between strength and indicators of body size and composition in Zapotec and Mexican American boys: all correlations are statistically significant (p < 0.001)

	Body mass	Stature	FFM	FM	% fat	Grip
Zapotec						
Age	0.76	0.79	0.72	0.83	0.80	0.67
Body mass		0.94	0.99	0.89	0.74	0.83
Stature			0.93	0.84	0.71	0.81
FFM				0.85	0.67	0.84
FM					0.95	0.70
% fat						0.56
Mexican American						
Age	0.79	0.86	0.80	0.45	0.34	0.79
Body mass		0.91	0.96	0.73	0.56	0.87
Stature			0.92	0.53	0.39	0.89
FFM				0.50	0.31	0.93
FM					0.94	0.41
% fat						0.25

Table 5. Partial correlations, controlling for age, body mass and stature, between grip strength and estimated body composition in Zapotec and Mexican American boys: all correlations are statistically significant (p < 0.001)

Body compartment	Zapotec	Mexican American
FFM	0.27	0.59
% fat	−0.25	−0.51
FM	−0.27	−0.59

their stature and body mass. These observations should be considered in the interpretation of body composition estimates in Zapotec boys.

The smaller body size and FFM of rural Zapotec boys compared to lower socioeconomic status Mexican American boys probably relates to differences in nutritional circumstances past and present. This is especially apparent in the gain of FFM between 9 and 14 years. On average, Zapotec boys gain 11.1 kg, while Mexican American boys gain 16.2 kg, a difference of 5.1 kg. Deficits in statural growth are similar, Zapotec boys gaining 22.9 cm between 9 and 14 years compared to a gain of 27.6 cm in Mexican American boys, a difference of 4.7 cm. The lesser increase in FFM and stature in Zapotec boys is likely related to their marginal nutritional circumstances. Estimated intakes of energy and protein among schoolchildren in the rural community approximate only 58 and 63%, respectively, of

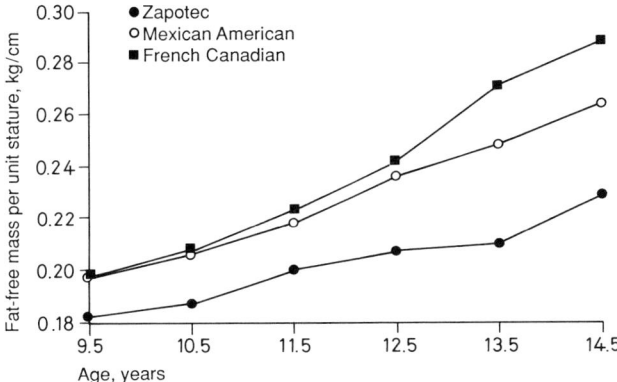

Fig. 6. Mean FFM per unit stature in Zapotec and Mexican American boys. Data for well-nourished French Canadian boys are included for comparison [Malina and Bouchard, unpubl. data]. All differences are significant ($p < 0.05$ 14 years, $p < 0.01$ 11 and 13 years, $p < 0.001$ all other ages).

recommended daily intakes for Mexico. Such intakes result in daily deficiencies which accumulate over time, resulting in smaller body size and reduced FFM.

The relationship between stature and FFM may indirectly provide an indication of changes in FFM associated with mild-to-moderate undernutrition. Stature is a cumulative measurement, representing the child's nutritional history. FFM is, to a large extent, a function of stature during childhood and adolesence [31]. The relationship between stature and FFM may be altered by chronically marginal nutritional circumstances. Correlations between stature and FFM in 9- to 14-year-old Zapotec and Mexican American boys are similar, 0.93 and 0.92, respectively (table 4), but FFM per unit stature is consistently lower in Zapotec boys (fig. 6). Data from a middle-class French Canadian population are included for comparison. Although the Mexican American boys are shorter and lighter and from a lower socioeconomic background compared to the French Canadian sample, FFM per unit stature is quite similar in the two samples, except at 13 and 14 years of age which is probably related to differential timing of the adolescent growth spurt.

Reduced FFM per unit stature in Zapotec boys would seem to suggest an altered relationship between FFM and stature with chronic protein-energy undernutrition. This is consistent with reduced intracellular water (body cell mass) per unit length or stature in undernourished Guatemalan children 2–8 years of age [32] and in free-living Australian aboriginal children 3–13 years of age [33]. Although FFM and body cell mass are not identical body composition concepts, the corresponding trends suggest a reduction in protein reserves per unit stature in chronically undernourished children. Muscle is a major protein reserve in the body and is a primary component of FFM. A functional consequence of less muscle mass is reduced muscular strength. Zapotec boys show absolutely lower grip strength than the better nourished Mexican American boys (fig. 3). More significantly, Zapotec boys have less strength per unit FFM (fig. 5), which may be suggestive of depleted protein reserves and altered composition of the FFM in children living under conditions of chronic protein-energy undernutrition. Although motivation is a factor in testing grip strength, age-specific correlations between first and second trials range from 0.82 to 0.97 in the Zapotec boys. Thus, it is reasonable to assume that the grip strength measurements are reliable and reflect maximal effort.

In contrast to FFM, differences in fatness between Zapotec and Mexican American boys are not marked. On average, the former accumulate only slightly less FM than the latter between 9 and 14 years, 2.6 vs. 3.5 kg, respectively, which is consistent with the small difference in relative fatness of the total sample of Zapotec and Mexican American boys respectively, 7.0 and 9.5%. This would suggest that protein rather than energy may be the more limiting nutrient in the diet of the rural indigenous community. This interpretation is consistent with the observations of Barac Neito et al. [12] on adult rural Colombian males. Changes in body composition with a moderate degree of chronic undernutrition were primarily related to a reduction in body cell mass (primarily muscle mass), while body fat was normal.

Comparative densitometric estimates of body composition in samples of indigenous Latin American children are not available. Viteri [11] used densitometry to estimate the body composition of young adult rural Guatemalan males. The relative fatness of Zapotec boys 9 through 14 years (7.0%) is quite similar to the estimate for very low socioeconomic status peasants from a very poor nutritional background who are engaged in active agricultural work (n = 16, mean age 18.6 years, mean fatness 8.1%). The mean stature and body mass of this sample of Guatemalan males (158.5 cm, 50.8 kg) are comparable to those of 17- to 19-year-old males in the Zapotec community (n = 76, mean age 18.6 years, mean stature 157.3 cm, mean body mass, 52.3 kg).

References

1 Waterlow JC, Cravioto J, Stephen JM: Protein malnutrition in man. Adv Prot Chem 1960;15:131–238.

2 Metcoff J: Biochemical effects of protein-calorie malnutrition in man. Ann Rev Med 1967;18:377–422.
3 Mendez J, Behrhorst C: The anthropometric characteristics of Indians and Urban Guatemalans. Hum Biol 1963;35:457–469.
4 Sabharwal KP, Morales S, Mendez J: Body measurements and creatinine excretion among upper and lower socioeconomic groups of girls in Guatemala. Hum Biol 1966;38:131–140.
5 Martorell R, Yarbrough C, Lechtig A, Delgado H, Klein RE: Upper arm anthropometric indicators of nutritional status. Am J Clin Nutr 1976;29:46–53.
6 Bogin BA, MacVean RB: Growth in height and weight of urban Guatemalan primary schoolchildren of low and high socioeconomic status. Hum Biol 1978;50:477–487.
7 Bogin BA, MacVean RB: Body composition and nutritional status of urban Guatemalan children of high and low socioeconomic status. Am J Phys Anthropol 1981;55:543–551.
8 Malina RM, Himes JH, Stepick CD, Gutierez Lopez F, Buschang PH: Growth of rural and urban children in the Valley of Oaxaca, Mexico. Am J Phys Anthropol 1981;55:269–280.
9 Spurr GB: Nutritional status and physical work capacity. Yearb Phys Anthropol 1983;26:1–35.
10 Spurr GB: Body size, physical work capacity, and productivity in hard work: Is bigger better? in Waterlow JC (ed): Linear Growth Retardation in Less Developed Countries. New York, Raven Press, 1988, pp 215–243.
11 Viteri FE: Consideration of the effect of nutrition on the body composition and physical working capacity of young Guatemalan adults; in Scrimshaw NS, Altschul AM (eds): Amino Acid Fortification of Protein Foods. Cambridge, MIT Press, 1971, pp 350–375.
12 Barac Nieto M, Spurr GB, Lotero GB, Maksud MG: Body composition in chronic undernutrition. Am J Clin Nutr 1978;31:23–40.
13 Malina RM: Physical activity and motor development/performance in populations nutritionally at risk; in Pollitt E, Amante P (eds): Energy Intake and Activity. New York, Liss, 1984, pp 285–302.
14 Malina RM: Motor development and performance of children and youth in undernourished populations; in Katch FI (ed): Sport, Health, and Nutrition. Champaign, Human Kinetics, 1986, pp 213–226.
15 Malina RM, Little BB: Body composition, strength, and motor performance in undernourished boys; in Binkhorst RA, Kemper HCG, Saris WHM (eds): Children and Exercise XI. Champaign, Human Kinetics, 1985, pp 293–300.
16 Malina RM, Selby HA, Swartz LJ: Estatura, peso y circunferencia del brazo en una muestra transversal de ninos Zapotecos de 6 a 14 anos. Anal Antropol 1972;9:143–155.
17 Malina RM, Selby HA, Buschang PH, Aronson WL: Growth status of school children in a rural Zapotec community in the Valley of Oaxaca, Mexico, in 1968 and 1978. Ann Hum Biol 1980;7:367–374.
18 Buschang PH, Malina RM: Growth in height and weight of mild-to-moderately undernourished Zapotec school children. Hum Biol 1983;55:587–597.
19 Buschang PH, Malina RM, Little BB: Linear growth of Zapotec schoolchildren: Growth status and yearly velocity for leg length and sitting height. Ann Hum Biol 1986;13:225–234.
20 Amdurer LRK: Nutrition in a Zapotec-speaking rural community, Oaxaca, Mexico; Master's thesis, University of Texas at Austin, 1978.
21 Pena Reyes M, Malina RM, Little BB, Buschang PH: Consumo de alimentos en una comunidad rural Zapoteca en el Valle de Oaxaca; in: Estudios de Antropologia

22 Malina RM: Growth and maturity profile of primary school children in the Valley of Oaxaca, Mexico. Garcia de Orta Ser Antropobiol 1983;2:153–157.
23 Malina RM, Selby HA, Buschang PH, Aronson WL, Wilkinson WG: Adult stature and age at menarche in Zapotec-speaking communities in the Valley of Oaxaca, Mexico, in a secular perspective. Am J Phys Anthropol 1983;60:437–449.
24 Zavaleta AN, Malina RM: Growth and body composition of Mexican American boys 9 through 14 years of age. Am J Phys Anthropol 1982;57:261–271.
25 Lohman TG: Skinfolds and body density and their relation to body fatness: A review. Hum Biol 1981;53:181–225.
26 Lohman TG: Applicability of body composition techniques and constants for children and youths. Exerc Sport Sci Rev 1986;14:325–357.
27 Malina RM: Growth of muscle tissue and muscle mass; in Falkner F, Tanner JM (eds): Human Growth, vol 2. Postnatal Growth. New York, Plenum, 1978, pp 272–294.
28 Himes JH: Bone growth and development in protein-calorie malnutrition. World Rev Nutr Dietet 1978;28:143–187.
29 Himes JH, Martorell R, Habicht JP, Yarbrough C, Malina RM, Klein RE: Patterns of cortical bone growth in moderately malnourished preschool children. Hum Biol 1975;47:337–350.
30 Himes JH, Malina RM, Stepick CD: Relationships between body size and second metacarpal dimensions in Oaxaca, Mexico, school children 6 to 14 years of age. Hum Biol 1976;48:677–692.
31 Forbes GB: Relation of lean body mass to height in children and adolescents. Pediatr Res 1972;6:32–37.
32 Cheek DB, Habicht JP, Berall J, Holt AB: Protein-calorie malnutrition and the significance of cell mass relative to body length. Am J Clin Nutr 1977;30:851–860.
33 Cheek DB, Graystone JE, Holt AB, Sutherland GC, Chopra SA, Spargo RM: Assessment of protein reserves (cellular mass) in aboriginal children. Am J Clin Nutr 1978; 31:1328–1333.
34 Malina RM, Hamill PVV, Lemeshow S: Selected body measurements of children 6–11 years, United States. Vital Health Statist 1973;series 11, No 123.

Robert M. Malina, PhD, Department of Kinesiology and Health Education, University of Texas, Austin, TX 78712 (USA)

Somatic Growth and Physical Performance in Canada

Roy J. Shephard

School of Physical and Health Education and Department of Preventive Medicine and Biostatistics, Faculty of Medicine, University of Toronto, Ont., Canada

Canada provides an interesting opportunity to study the respective influences of environment and inheritance upon growth and physical performance. There remain relatively pure samples of the indigenous peoples, both North American Indians and Inuit, with growth, developmental and performance data recorded at various stages in their acculturation to a North American lifestyle. In the central part of the Province of Quebec, one can find equally pure samples of the original French Canadian settlers, while in Ontario and the western provinces the typical Scottish and English settlers are rapidly giving place to a multicultural North American 'melting pot'. The data have particular value in that information on growth, nutrition and physical performance is sometimes available for the same populations, while government funding has allowed the collection of information on large and representative samples of these populations, using well-standardised and internationally accepted methodology [Shephard, 1978a, 1986].

For the purposes of this chapter, we shall consider growth statistics from a substantial number of Canadian surveys, and will then focus specifically upon fitness and performance data obtained in five national and two regional surveys, one of the latter being conducted in French Canada, and the other among the Inuit of the circumpolar community.

Methods

Sample Selection

When interpreting studies in cultural anthropology, one major source of difficulty has been the failure of investigators to select a representative sample of the population of interest; the subjects are frequently small groups of conveniently available volunteers [Shephard, 1978a]. The Canadian data are particularly satisfactory in this regard. All of the major national studies to be discussed in detail sought the guidance of Statistics Canada in order to obtain not only large but also representative samples of the Canadian population. For the measurements of working capacity and physical performance undertaken by the Canadian Association of Physical and Health Education [Howell and MacNab, 1968; Gauthier et al., 1983; Hayden and Yuhasz, 1966; Gauthier, 1980], stratification of the population was based on a series of appropriately distributed schools, while in the Canada Fitness Survey [1983], clusters of subjects in numbers proportional to the square root of the population of each province were tested by some 80 appropriately distributed measurement teams. In the Trois Rivières regional study [Shephard, 1982a], entire school classes were tested in an urban and a rural setting, while in circumpolar investigations of the Canadian Inuit, all willing children in the Igloolik region (some 72% of the total population) were examined [Rode and Shephard, 1973, 1984].

Age of Subjects

A second critical variable is the specification of subject age. All of the national surveys in Canada reported the age at the child's last birthday; given the large size of the samples tested, the effective age of those reported as 7 years was thus 7.5 years. Most of the regional surveys, including the Trois Rivières study and the circumpolar investigations, have reported the actual age of the child; in Trois Rivières, all data were collected annually, within 2 weeks of each child's birthday.

Cross-Sectional versus Longitudinal Design

In an era of secular change, a further important issue is the use of cross-sectional rather than longitudinal data. All of the five national surveys in Canada were cross-sectional in type, but fortunately both the fitness and the performance measurements were repeated after an interval of 15 years, applying an identical methodology for both subject selection and testing. The Trois Rivières regional study followed cohorts of experimental and control students for a 6-year period, while the circumpolar study was cross-sectional in design, but observations were repeated by the same investigators on the same community after the lapse of 1 and 10 years.

Measurement Techniques

Finally, there is the issue of measurement techniques. Scores for most indices of working capacity and physical performance are heavily influenced by the test procedures that are used. In all of the Canadian studies to be considered in detail, methods were carefully standardized; moreover, the observers received centralized training and followed detailed instructions published in the reports that are cited. In general, the techniques that were used conformed to the recommendations of the International Biological Programme [Weiner and Lourie, 1981]. Some authors have argued that data should be aligned in terms of peak height velocity or some similar index of maturity. Unfortunately, this presupposes a longitudinal study, with several accurate measurements of standing height per year. Such information was not available for most of the Canadian data to be discussed, and accordingly the simpler expedient of chronological age has been used here.

Results

Height

Canadian national data are available from 1954 [Pett and Ogilvie, 1956], 1979 [Gauthier, 1980] and 1981 [Canada Fitness Survey, 1983] (table

1). Inuit on the east coast of Hudson Bay were examined in 1965 by Partington and Roberts [1969], while Inuit of the Melville Peninsula (Igloolik) and of Northern Quebec (Fort Chimo) were studied in 1970/1971. Observations were repeated at Igloolik in 1980/1981 [Rode and Shephard, 1973, 1984; Demirjian et al., 1976]. Cree Indians of the James Bay region and relatively prosperous Mohawk Indians living on a reserve near Kingston, Ont. were measured in 1965 by Partington and Roberts [1969]. In 1971, French Canadian children living in Montreal were evaluated by Demirjian et al. [1972], while those of the Trois Rivières region were studied longitudinally from 1970 to 1975 [Shephard, 1982]. English Canadian samples include 1943/1945 data for the Ottawa region [Hopkins, 1947], 1968 data for London, Ont. [Stennett and Cram, 1969], 1970 data for Halifax [Welch et al., 1971], and longitudinal data from the 1970s for Saskatoon [Demirjian et al., 1976].

Insofar as can be judged from what is all cross-sectional data, the age of peak height velocity in the national samples appeared to be a few months earlier in the 1979 and 1981 surveys than in 1954. There was also a pronounced secular trend to increase of stature over the intervening 25–27 years; for instance, in prepubertal boys (aged 9 years), the increase amounted to 4.0–4.5 cm (1.63 cm/decade), while at 15 years (after the adolescent growth spurt) it was 5.4–5.8 cm (2.15 cm/decade). Corresponding figures for the girls were 5.2–5.6 cm (2.08 cm/decade) and 3.5–3.8 cm (1.40 cm/decade). All of these figures seem greater than for many developed nations.

All figures for the Inuit showed a much shorter average stature, with little difference between subjects from the Igloolik and Fort Chimo regions; for example, at 9 and 15 years the 1970/1971 figures for the Inuit boys were 4.6 and 9.8 cm below the 1954 national standards, while in the girls the deficits were 8.0 and 7.4 cm. Analysis of adult heights at first suggested a strong secular trend to increase of stature among Canadian Inuit, mirroring that reported for the Lapps [Skrobak-Kaczynski and Lewin, 1976] but further investigation of the Igloolik sample showed little increase of stature in either children or young adults from 1970/1971 to 1980/1981, while many of the older adults showed a decrease in height of some 2 cm over the decade; it was suggested that a part of the current steep gradient of height among adult Inuit, although mimicking a rapid secular trend, was really an artefact of spinal trauma due to prolonged high-speed snowmobile journeys over rough ice [Shephard et al., 1985].

The Indian children of 1965 were generally taller if living in the Mohawk reservation near Kingston than if located in the more Northerly Cree reserve, the average difference of stature amounting to some 2.3 cm. Allowing for the age grouping adopted by Partington and Roberts [1969],

Table 1a. Height of Canadian boys (cm): the two ages encompassing peak height velocity are *italicized* for each sample.

Age years	National samples			Inuit				Indian		French Canadian		English Canadian				
	1954[1]	1979[2]	1981[3]	1965[4]	1970/1971[5,6]		1980/1981[7]	1965[4]		1969[8]/1975[9]	1970/	1945[10]		1969[11]	1971[12]	1976[5]
					I	F		C	M			H	L			
1	*	—	—	—	—	—	—	—	—	—	—	—	—	—	—	—
2	86.6	—	—	—	—	—	—	—	—	—	—	—	—	—	—	—
3	91.7	—	—	—	—	—	—	—	—	—	—	—	—	—	—	—
4	99.6	—	—	*x	—	—	—	—	—	—	—	—	—	—	—	—
5	106.9	*	—	109.7	x	—	x	*	—	—	—	—	—	—	—	—
6	113.3	119.8	*	112.8	116.5	117.0	—	117.3	115.8	113.9	114.1	—	—	114.3	—	—
7	118.9	124.7	123.5	118.6	121.2	121.6	—	123.1	121.9	119.7	119.9	121.1	118.1	120.7	—	122.3
8	124.5	130.0	129.3	124.0	125.7	126.1	126	128.3	128.5	125.5	125.4	126.5	122.8	125.7	126.0	127.6
9	130.3	134.3	134.8	128.8	128.6	133.8	129	131.7	135.9	130.2	130.9	132.4	129.6	130.8	—	133.1
10	135.4	140.7	140.2	133.4	133.1	134.0	133	136.1	141.7	135.6	135.4	136.9	134.5	137.2	—	138.5
11	140.5	145.0	145.0	137.4	138.5	142.3	143	141.0	143.5	140.0	140.4	141.1	138.0	142.2	141.8	143.4
12	145.3	*151.4*	151.3	140.2	144.8	142.0	148	*145.9*	*150.6*	*145.4*	144.7	—	—	147.3	—	148.8
13	*150.1*	*158.3*	*157.8*	—	*144.8*	*154.4*	151	*152.9*	*159.0*	*152.8*	—	—	—	*153.7*	—	*154.9*
14	*150.8*	164.6	*167.1*	—	149.1	157.5	158	159.6	—	158.4	—	—	—	*161.3*	—	*162.1*
15	165.1	170.5	170.9	—	156.3	—	—	163.7	—	166.3	—	—	—	167.6	—	169.0
16	171.7	173.7	173.3	—	—	—	162	—	—	170.0	—	—	—	170.2	—	—
17		174.4	175.1	—	—	—	163	—	—	171.8	—	—	—	174.0	—	—
18	173.2	—	175.4	—	164.3	—	163	—	—	—	—	—	—	175.3	—	—
19		—	175.6	—	—	—	163	—	—	—	—	—	—	—	—	—

* Age group (i.e. 7 years = 7.00 – 7.99 years). H = High economic status; L = low economic status; C = Cree; M = Mohawk; x = interpolated values; I = Igloolik; F = Fort Chimo.

[1] Pett and Ogilvie [1956].
[2] Gauthier [1980].
[3] Canada Fitness Survey [1983].
[4] Partington and Roberts [1969].
[5] Demirjian et al. [1976].
[6] Rode and Shephard [1973].
[7] Rode and Shephard [1984].
[8] Demirjian et al. [1972].
[9] Shephard [1982].
[10] Hopkins [1947].
[11] Stenett and Cram [1969].
[12] Welch et al. [1971].

Table 1b. Height of Canadian girls (cm): the ages encompassing peak height velocity are *italicized* for each sample

Age years	National samples			Inuit		1980/ 1981[7]	Indian 1965[4]		French Canadian		English Canadian				
	1954[1]	1979[2]	1981[3]	1965[4]	1970/ 1971[5,6]				1969[8]	1970/ 1975[9]	1945[10]		1969[11]	1971[12]	1976[5]
							C	M			H	L			
1	*	*	—	—	—	—	—	—	—	—	—	—	—	—	—
2	86.6	—	—	—	—	—	—	—	—	—	—	—	—	—	—
3	91.7	—	—	—	—	—	—	—	—	—	—	—	—	—	—
4	99.6	—	—	*x	—	—	—	—	—	—	—	—	—	—	—
5	106.9	*	—	107.7	—	—	*	*	—	—	—	—	—	—	—
6	113.3	119.6	*	111.5	x	—	116.8	115.1	112.7	112.8	—	—	114.3	—	—
7	118.4	123.6	124.2	117.1	112.0	—	123.3	119.4	118.5	118.7	120.9	119.0	119.4	—	121.2
8	124.2	129.2	129.2	118.9	117.6	—	124.5	130.0	124.3	124.7	125.5	123.0	124.5	125.0	126.7
9	129.0	134.2	134.8	124.0	121.0	x	132.0	133.4	129.0	130.0	131.2	128.4	130.8	—	132.5
10	134.9	139.9	141.2	132.6	127.4	131	*136.8*	*138.2*	135.4	134.8	136.8	134.1	*135.9*	*142.7*	138.3
11	*140.5*	*146.1*	*145.9*	*137.2*	*137.0*	137	*134.4*	*149.1*	*141.5*	*140.9*	*141.6*	*138.1*	*146.1*	*150.0*	*143.7*
12	*148.1*	*153.2*	*152.8*	*144.0*	*139.9*	142	*149.1*	*153.2*	*147.6*	*148.2*	—	*148.6*	—	*156.7*	
13	*153.7*	158.1	158.4	*147.6*	*145.9*	147	155.1	157.7	153.9	—	—	154.9	—	161.0	
14	155.7	160.4	162.0	150.0	150.4	148	156.1	—	158.2	—	—	158.8	—	163.9	
15	158.0	161.5	161.8	153.4	150.6	151	159.6	—	160.6	—	—	160.0	—	—	
16	159.0	163.1	162.0	—	—	152	—	—	160.3	—	—	160.0	—	—	
17		163.4	161.9	—	—	153	—	—	—	—	—	160.0	—	—	
18	159.3	—	162.6	156.0	—	153	—	—	—	—	—	162.6	—	—	
19		—	162.3	—	—	153	—	—	—	—	—	—	—	—	

Footnotes as in table 1a.

Table 2a. Body mass (kg) of Canadian boys and girls

Age years	National samples					Inuit					Indian		French Canadian		English Canadian			
	1954[1]	1966[2]	1979[3]	1981[4]	1981[5]	1965[6]	1970/1971[7,8]	1971[9] I	1971[9] F	1980/1981[10]	1965[6] C	1965[6] M	1969[11]	1970/1975[12]	1945[13]	1969[14]	1971[15]	1970/1975[9]
1	*	—	—	—	—	—	—	—	—	—	—	—	—	—	—	—	—	—
2	13.2	—	—	—	—	—	—	—	—	—	—	—	—	—	—	—	—	—
3	14.2	—	—	—	—	—	—	—	—	—	—	—	—	—	—	—	—	—
4	16.5	—	—	—	—	*	—	—	—	—	—	*	—	—	—	—	—	—
5	17.8	—	*	—	—	20.5	x	—	—	—	—	—	—	—	—	—	—	—
6	20.2	*	22.9	*	*	21.8	21.6	—	—	—	23.3	23.6	20.2	20.6	—	20.5	—	—
7	22.6	25.7	25.3	24.5	25.4	25.3	23.9	22.7	22.9	—	25.3	26.7	22.5	22.6	23.5	22.8	—	23.2
8	25.4	27.7	28.3	28.2	28.3	26.7	26.4	24.8	24.9	x	27.2	30.2	24.8	25.1	26.1	24.7	26.1	25.7
9	28.2	30.5	30.8	31.1	31.0	29.8	28.8	27.5	26.6	29	29.7	31.4	27.7	28.1	29.3	27.5	—	28.7
10	30.8	33.7	35.7	34.9	33.7	32.3	31.2	30.2	31.5	29	32.5	38.7	30.3	30.5	31.7	30.2	—	31.9
11	33.9	37.7	38.0	38.6	39.7	34.1	34.2	32.3	29.7	31	34.4	39.0	33.2	33.5	34.5	31.4	—	35.3
12	37.0	42.2	43.0	42.6	43.3	36.1	38.3	35.2	35.3	37	38.4	40.9	38.6	36.8	—	34.5	36.0	39.1
13	41.3	47.1	48.3	47.4	49.7	—	42.3	41.1	38.2	43	42.1	53.6	44.9	—	—	32.9	—	43.8
14	47.9	53.1	54.7	56.8	54.5	—	47.7	43.6	46.2	43	49.2	—	47.9	—	—	39.1	—	49.6
15	53.9	58.2	60.9	63.2	60.1	—	53.6	51.1	50.2	50	51.8	—	55.5	—	—	43.2	—	56.0
16		64.0	63.9	64.1	65.8	—	—	53	—	55	—	—	58.6	—	—	51.4	—	—
17	61.4	65.4	66.2	66.9	66.6	—	60.6	56	—	59	—	—	59.3	—	—	56.4	—	—
18		—	—	69.6	—	—	—	62	—	63	—	—	—	—	—	60.5	—	—
19	63.7	—	—	70.2	—	—	—	61	—	63	—	—	—	—	—	65.0	—	—
																66.4		

*Age group (i.e. 7 years = 7.00 – 7.99 years). x = Interpolated; C = Cree; M = Mohawk; I = Igloolik; F = Fort Chimo; H = high economic status; L = low economic status.

[1] Pett and Ogilvie [1956]: subjects wearing indoor clothing.
[2] Howell and MacNab [1968].
[3] Gauthier [1980].
[4] Canada Fitness Survey [1983].
[5] Gauthier et al. [1983].
[6] Partington and Roberts [1969].
[7] Mazess and Mather [1975].
[8] Rode and Shephard [1973].
[9] Demirjian et al. [1976].
[10] Rode and Shephard [1984].
[11] Demirjian et al. [1972].
[12] Shephard [1982].
[13] Hopkins [1947].
[14] Stennett and Cram [1969].
[15] Welch et al. [1971].

Table 2b. Body mass of Canadian girls (kg)

Age years	National samples						Inuit				Indian		French Canadian		English Canadian			
	1954[1]	1966[2]	1979[3]	1981[4]	1981[5]	1965[6]	1970/1971[7,8]	1976[9]	1980/1981[10]	1965[6] C	M	1969[11]	1970/1975[12]	1945[13] H	L	1969[14]	1971[15]	1970/1975[9]
1	*	–	–	–	–	–	–	–	–	–	–	–	–	–	–	–	–	–
2	13.2	–	–	–	–	–	–	–	–	–	–	–	–	–	–	–	–	–
3	14.2	–	–	–	–	–	–	–	–	–	–	–	–	–	–	–	–	–
4	16.5	–	–	–	–	*x	–	–	–	–	–	–	–	–	–	–	–	–
5	17.8	–	*	–	–	20.0	–	–	–	–	–	–	–	–	–	–	–	–
6	20.2	*	22.9	–	*	21.8	20.6	x	–	22.8	21.8	19.9	19.5	–	–	20.0	–	–
7	21.9	24.0	24.7	25.3	25.0	23.4	32.8	21.7	–	25.2	26.9	21.6	21.7	23.6	22.5	22.3	–	22.8
8	24.5	27.2	27.6	28.2	29.0	23.9	25.1	23.8	–	26.3	30.0	25.5	24.7	25.9	24.4	24.5	25.6	25.3
9	27.2	30.1	31.2	30.6	31.3	–	27.4	26.0	x	29.8	35.0	27.2	27.4	28.6	27.0	27.7	–	28.5
10	30.3	33.5	34.8	36.2	35.2	–	29.7	30.1	34	32.5	35.5	30.9	30.3	31.4	30.1	30.9	–	32.0
11	33.3	36.2	38.9	37.7	39.0	–	32.2	34.2	36	36.5	43.9	34.5	34.1	35.2	32.1	35.9	36.3	35.9
12	40.5	43.1	45.6	45.0	43.4	39.5	39.2	37.5	37	42.5	43.6	38.8	38.4	–	–	40.0	–	41.1
13	46.4	49.8	50.3	49.6	51.5	–	44.5	43.3	43	46.1	48.7	44.9	–	–	–	45.5	–	45.8
14	47.4	52.3	52.4	55.0	50.4	47.3	49.8	48.5	45	50.7	–	49.5	–	–	–	50.5	–	51.6
15	50.0	54.1	54.2	54.7	58.1	–	52.2	50.8	47	50.9	–	52.1	–	–	–	52.7	–	51.6
16	54.7	56.1	56.3	58.2	–	53.4	53	50	–	–	51.4	–	–	–	55.0	–	–
17	54.6	56.7	56.1	56.9	56.8	–	53.4	54	51	–	–	–	–	–	–	–	–	–
18	–	–	58.0	–	–	–	53	51	–	–	–	–	–	–	–	–	–
19	54.6	–	–	58.8	–	–	–	52	52	–	–	–	–	–	–	–	–	–

Footnotes as in table 2a.

the height of the Mohawk group was intermediate between that of the 1954 and 1979 national samples, corresponding to about 0.3 cm with figures predicted from a linear secular trend.

Statistics for the standing height of French Canadian children showed a close agreement between the Montreal and Trois Rivières regional samples, both groups being 2–3 cm shorter than would have been predicted from secular trends in the national population; a difference of this order had previously been foreshadowed by a regional breakdown of national data collected in 1954 [Demirjian et al., 1972].

Regional data from English Canada are generally unremarkable, although the figures of Hopkins [1947] for the city of Ottawa illustrate that towards the end of World War II, socioeconomic factors were associated with a 2–3-cm difference of stature among this population.

Body Mass

The mean figures from the national data for 1954 (table 2) correspond closely with the median figures suggested by the US National Center for Health Statistics [1976] for children of comparable age. However, figures reported from recent surveys (1979–1981) exceed these standards by some 3 kg at age 9 years, both in boys and in girls. It does not seem possible to explain the increase of body mass simply in terms of the secular trend to an increase of stature. In 1954, the somewhat dated 'weight for height standards' [Baldwin, 1925, as cited by Jelliffe, 1966] were exceeded by 1.5 kg in 15-year-old boys, and 0.1 kg in 15-year-old girls, but by the 1981 Canada Fitness Survey discrepancies at the same age had increased to 4.9 and 1.9 kg. Likewise, the index ($W/H^3 \times 100$) for 15-year-old subjects had increased from 1.198 in the boys and 1.268 in the girls of 1954 to 1.266 and 1.291 for the two sexes in 1981.

The Inuit of the Igloolik region were substantially lighter than the general Canadian sample, even in the 1980/1981 survey, but when figures were expressed relative to the cube of height, the relative body mass was seen to be greater than in the general population; at age 9 years, the figures for Inuit boys and girls were 1.460 and 1.806, compared with 1.270 and 1.244 in the Canada Fitness survey, while at age 15 years, the corresponding values were 1.268 and 1.365 vs 1.266 and 1.291. The differences reflect both a stocky frame and greater muscularity in the unacculturated Inuit.

The Mohawk Indians were substantially heavier than the Cree; at 9 years of age, the boys from the two reserves showed a difference of 1.7 kg, while in the girls the difference was 5.2 kg. When results were calculated relative to the cube of height, the corresponding Indices for the Cree were above the current national average in both boys (1.300) and girls (1.296); the corresponding indices for the Mohawks were 1.251 and 1.474.

French Canadian children were lighter than the national average, whether living in Montreal or Trois Rivières, but this was again largely due to their short stature. Taking the figures of Demirjian et al. [1972] and relating body mass to the cube of height, the indices were very normal values of 1.255 and 1.267 at 9 years of age and 1.207 and 1.258 at 15 years of age for boys and girls, respectively.

Among the English Canadian regional samples, the data of Hopkins [1947] showed a substantially greater body mass in children from the upper socioeconomic category. Expressing values at age 9 years in terms of height cubed, indices were nevertheless relatively similar for boys and girls of high (1.262, 1.266) and low (1.263, 1.275) socioeconomic status.

Subcutaneous Fat

A comparison of national samples for 1954 and 1981 suggests that there was a substantial increase in the average thickness of triceps skinfolds over this interval (table 3). In 1970/1971, Inuit children from Igloolik were thinner than the national sample of 1954, but they also showed a substantial increase of average skinfold readings during the decade from 1970/1971 to 1980/1981; over this same period, the Igloolik community underwent a rapid acculturation from a society where 30–40% of food was obtained by hunting and trapping to a much more unbanized lifestyle, with most of the food imported from southern Canada. Both French and English Canadian data from 1968–1969 seem intermediate between the values observed in the 1954 and 1981 national surveys.

Unfortunately, there have been no large-scale surveys of Canadians which have made more direct estimates of the percentage of body fat. However, where several skinfolds have been measured, it is possible to use standard age-specific formulae to predict the percentage of body fat [Shephard et al., 1969] with moderate accuracy. Application of this prediction technique to the French Canadian children studied in the Trois Rivières region (fig. 1) shows a relatively normal distribution in the predicted percentage of body fat at the age of 6 or 7 years, but a substantial skewing of the same data by the age of 12 years. This seems to imply that a proportion of the sample have become obese; in the boys, there is an associated increase in the mean predicted percentage of fat from 13% to more than 16%, while in the girls the increase is from 19% to more than 22%.

Physical Working Capacity

The national samples studied by the Canadian Association for Health, Physical Education and Recreation (CAHPER; table 4) show a 10–20% increase of the physical working capacity at a heart rate of 170 beats/min

Table 3. Skinfold thickness in Canadian children

Age years	Boys								Girls							
	National		Inuit		French Canadian 1969[5]		English Canadian		National		Inuit		French Canadian 1969[5]		English Canadian 1968[6]
	1954[1]	1981[2]	1970/1971[3]	1980/1981[4]			1968[5]	1970/1975[7]	1954[1]	1981[2]	1970/1971[3]	1980/1981[4]			
	*T	*T	T+S+Si	T	T	S	T+S+Si		*T	*T	T+S+Si	T	T	S	T+S+Si
1	9.4	—	x	x	—	—	—		9.4	—	x	—	—	—	—
2	9.1	—	—	—	—	—	—		9.1	—	—	—	—	—	—
3	8.5	—	—	—	—	—	—		8.5	—	—	—	—	—	—
4	7.7	—	—	—	—	—	—		7.7	—	—	—	—	—	—
5	7.7	—	—	—	—	—	—		7.6	—	—	—	—	—	—
6	7.6	—	—	—	8.0	5.9	—		8.0	11.0	—	—	9.6	6.9	—
7	6.4	9.1	—	—	8.1	5.8	—		8.5	12.2	—	—	9.6	6.7	—
8	6.4	9.7	—	—	8.1	6.2	—		8.5	11.8	4.3	—	10.4	8.1	—
9	6.4	9.8	3.7	4.7	8.7	6.3	—		8.6	13.0	5.7	—	11.1	7.8	—
10	6.7	10.5	4.0	5.3	8.1	6.3	7.6		9.1	12.2	5.3	8.4	10.8	7.6	10.2
11	7.0	11.3	4.3	5.7	9.2	6.9	7.6		8.9	12.8	6.7	7.3	11.7	8.1	10.2
12	6.9	10.8	5.3	6.7	9.9	8.1	8.0		10.0	13.5	7.7	7.3	11.5	8.7	
13	6.7	9.6	5.0	6.7	10.1	8.3	8.0		10.7	15.0	8.0	8.0	12.5	9.6	
14	6.2	9.4	5.0	5.7	9.0	7.9	—		10.2	14.6	9.7	8.0	13.3	10.6	—
15	5.4	9.1	4.7	5.7	9.5	8.5	—		13.1	16.2	10.7	11.3	14.8	11.1	—
16	5.2	9.0	5.0	5.7	8.0	8.4	—		13.1	16.3	10.7	10.7	13.8	10.8	—
17	5.2	8.9	5.7	6.7	8.1	8.6	—		13.1	16.3	11.0	11.7	—	—	—
18	5.3	9.5	6.0	7.3	—	—	—		13.1	17.3	9.3	11.0	—	—	—
19	5.3	9.3	6.0	8.0	—	—	—					11.1			

* Age at last birthday. T = Triceps fold; S = subscapular fold; Si = suprailiac fold; x = interpolated.
[1] Pett and Ogilvie [1956].
[2] Canada Fitness Survey [1983].
[3] Rode and Shephard [1973].
[4] Rode and Shephard [1984].
[5] Jenicek and Demirjian [1972].
[6] Shephard et al. [1969].
[7] Bailey et al. [1974].

Table 4. Physical working capacity at heart rate of 170 beats/min (W, W/kg)

Age years	Boys							Girls						
	National 1966[1]		1981[2]		English Canadian 1968[3]	French Canadian 1970/1975[4]		National 1966[1]		1981[2]		English Canadian 1968[3]	French Canadian 1970/1975[4]
	W	W/kg	W	W/kg	W	W/kg		W	W/kg	W	W/kg	W	W/kg
6	–	–	–	–	–	2.23		–	–	–	–	–	1.53
7	51.1	2.00	54.9	2.17	–	2.22		47.2	1.60	46.0	1.89	–	1.82
8	58.5	2.12	62.2	2.23	–	2.27		47.5	1.78	51.8	1.81	–	1.90
9	64.1	2.12	71.6	2.35	–	2.62		51.1	1.71	60.7	1.97	–	2.02
10	71.2	2.13	76.0	2.29	53.2	2.57		56.2	1.69	68.1	1.96	47.9	2.12
11	82.3	2.20	91.7	2.36	95.0	2.68		60.2	1.68	75.5	1.97	70.4	2.13
12	92.3	2.23	94.2	2.22	–	2.77		69.6	1.65	78.7	1.83	–	2.03
13	109.2	2.33	111.4	2.28	–	–		75.1	1.53	92.5	1.84	–	–
14	121.4	2.31	133.7	2.47	133.0	–		72.6	1.42	91.1	1.82	81.2	–
15	123.3	2.14	146.5	2.47	158.0	–		73.9	1.35	104.5	1.83	98.8	–
16	142.1	2.24	166.6	2.55	–	–		76.6	1.41	106.9	1.85	–	–
17	143.9	2.23	158.7	2.40	–	–		79.5	1.42	106.9	1.88	–	–

[1] Howell and MacNab [1968].
[2] Gauthier et al. [1983].
[3] Alderman [1969].
[4] Shephard [1982a].

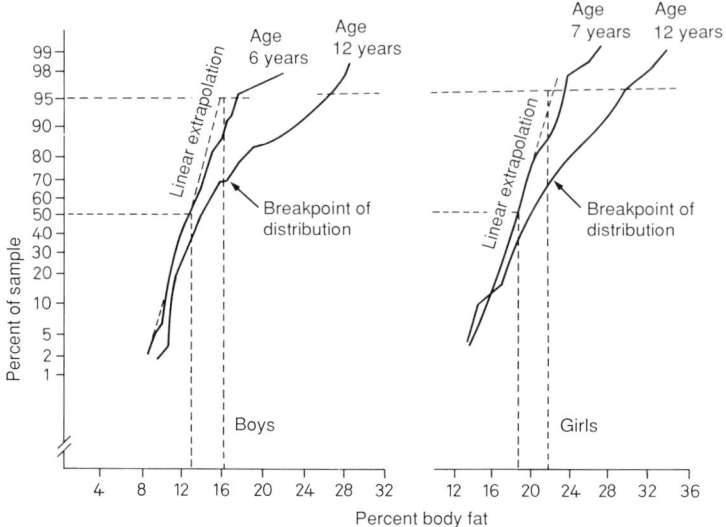

Fig. 1. Probit plot showing distribution of body fat in population of students enrolled in the Trois Rivières longitudinal study. Source: Shephard [1978b].

from 1966 to 1981 [Howell and MacNab, 1968; Gauthier et al., 1983], the gain in score being evident whether the data are expressed in absolute units of power output (W) or relative to body mass (W/kg). The Toronto segment of the CAHPER sample was retested in an air-conditioned laboratory [Shephard et al., 1969], and the values observed at retest were also some 10% higher than those reported as the national average. Other values for a limited sample of English Canadian children, as reported by Alderman [1969] show substantial variability relative to the national standards. The figures for the French Canadian students at Trois Rivières (fig. 2) were relatively high [Shephard, 1982], in part because half of the sample tested had been undertaking an hour of additional physical education per day as part of a long-term training project.

Maximum Oxygen Intake

Unfortunately, there have been no direct measurements of maximum oxygen intake on national samples of Canadian children (table 5). The Canada Fitness Survey used a simple step test from which some rather crude estimates of maximum oxygen intake can be formed [Shephard, 1986]. Some estimate of aerobic power can also be derived by applying the Åstrand nomogram to the PWC_{170} data obtained by CAHPER in 1966 and 1981, although it is likely that a prediction based on a single submaximal

Table 5. Maximum oxygen intake of Canadian children (ml/kg · min)

Age years	National			Inuit		French Canadian		English Canadian			
	1981[1]	1966[2]	1981[3]	1970/ 1971[4]	1980/ 1981[5]	1974[6]	1974[7]	1967[8]	1968[9]	1973[10]	1970/ 1975[11]
Boys											
6	*E	S	S	S	S	—	—	—	—	—	—
7				x	x	—	—	—	—	—	—
8	49.9	38.4	41.3	—	—	—	—	—	—	—	56.4
9				69	58	—	—	—	—	—	59.5
10				65	64	46.7		44		52.5	56.9
11	50.4	40.2	41.8	65	65	—	54.3	—	47.4	—	56.3
12				70	60	—		47		—	56.6
13				69	58	—		—		52.9	55.1
14	50.7	42.3	43.3	66	59	—	—	49	—	—	54.6
15				65	66	—	—	—	—	—	52.6
16				65	62	—	—	52	—	56.6	—
17	49.3	40.9	43.7	64	55	—	—	—	—	—	—
18				64	53	—	—	55	—	—	—
19				63	50	—	—	—	—	—	—
Girls											
6	*E	—	—	S	S	—	—	52	—	—	—
7				x	x	—	—	—	—	—	—
8	42.7	33.8	36.8	—	—	—	—	49	—	—	—
9				54	—	—	—	—	—	—	—
10				58	47	38.1		40	36.9	—	—
11	42.8	33.4	37.5	57	48	—	42.4	—	—	—	—
12				53	50	—		42	—	—	—
13				51	49	—	—	—	—	—	—
14	42.5	30.0	35.9	51	45	—	—	38	—	40.5	—
15				50	45	—	—	—	—	—	—
16				49	45	—	—	39	—	—	—
17	40.9	29.0	36.8	50	45	—	—	—	—	—	—
18				51	54	—	—	44	—	—	—
19				52	45	—	—	—	—	—	—

* Completed years; E = estimate from Canadian Home Fitness Test [Shephard, 1986]; S = prediction from submaximal test.
[1] Canada Fitness Survey [1983].
[2] Howell and McNab [1968].
[3] Gauthier et al. [1983].
[4] Rode and Shephard [1973].
[5] Rode and Shephard [1984].
[6] Larivière et al. [1974].
[7] Shephard et al. [1974].
[8] Cumming [1967].
9 Shephard et al. [1969].
[10] Cunningham and Eynon [1973].
[11] Bailey et al. [1978].

cycle ergometer test without preliminary habituation of the subjects underestimates the true maximum oxygen intake by up to 15% in this age group [Shephard, 1978].

Data obtained on the Inuit of the Igloolik region during 1970/1971 showed a substantially higher maximum oxygen intake than either our estimates for the national sample or most figures from other countries. Although the majority of results for the Inuit were predicted from heart rate and oxygen consumption during a progressive submaximal step test, the unusually high values were confirmed by direct measurement of maxima on a proportion of the sample. By 1980/1981, much of this advantage had disappeared, particularly in the girls and the older boys [Rode and Shephard, 1984].

Somewhat disparate scores were obtained on two relatively small samples of French Canadian students from the Trois Rivières region who were tested in 1974 [Larivière et al., 1974; Shephard et al., 1974]. However, the average of the two results did not differ substantially from the national figures. Values for the much more substantial Trois Rivières longitudinal study [Shephard, 1982] tended towards the higher of these two figures, as might be anticipated from the involvement of a half of the students in a regular endurance training programme (fig. 2). The Trois Rivières experiment established that an additional hour per day of physical education progressively increased the directly measured maximum oxygen intake of primary school students by 10–15% relative to other students of similar age who were enrolled in the normal French Canadian school programme which offered only 40 min of physical education per week, taught by a nonspecialist.

Most regional studies from English Canada have involved rather small samples of students. The report of Shephard et al. [1969] has particular interest in that (i) it describes the conclusions of the IBP working party on the standardization of methodology for the evaluation of children, and (ii) the participating subjects represent the Toronto component of the CAHPER national study described by Howell and MacNab [1968]. Cunningham and Eynon [1973] studied swimmers, while the students examined by Bailey et al. [1978] also had above average involvement in physical activity; as in the Trois Rivières study, scores for these two active samples were some 10% above the national average.

Physical Performance Test Scores

Field performance test scores were obtained on a representative sample of Canadian children in 1964 [Hayden and Yuhasz, 1966] and again in 1979 [Gauthier, 1980]. Results were obtained on students of all ages from 7 to 17 years in 1964, and from 6 to 17 years in 1979. During the second

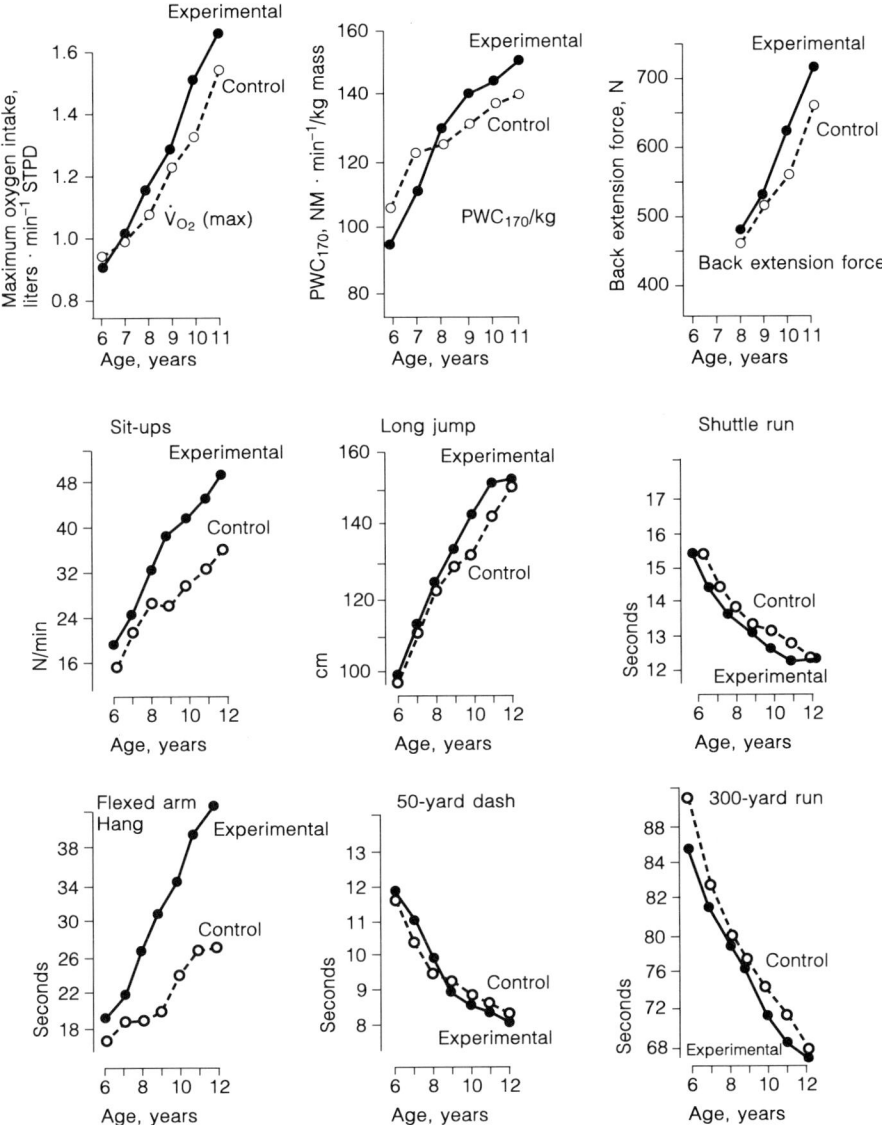

Fig. 2. Some measures of physical fitness and performance test scores for students enrolled in the Trois Rivières longitudinal study. Experimental group received five additional hours of physical education per week. Source: Shephard [1982b].

survey, details of protocol were modified for some tests. In particular, the shuttle distance run was increased from 10 yards (9.2 m) to 10 m, and the length of the sprint run was raised from 50 yards (45.9 m) to 50 m; assuming unchanged performance, this metrification should have increased times by about 9.4%.

The body mass of the children increased from 1964 to 1979, and although heights were not measured in 1964, it is probably from other figures for Canada that there was some increase of stature over this same interval; unfortunately, test scores are highly susceptible to both height and body mass [Cumming, 1971], and also improve with test learning. Nevertheless, there is plainly a secular trend to an increase of strength, particularly in the older girls (who showed large increases in scores for speed sit-ups, standing broad jump and flexed arm hang). Comparisons with results from other nations are difficult, because of differences in both body size and test protocol. The most closely comparable tests are those conducted by the American Association for Health, Physical Education, Recreation and Dance [AAHPER, 1958, 1965, 1976]. In the US, also, there has been a secular trend to an improvement of performance test scores. However, the Canadian girls of both 1964 and 1979 (table 6) outperformed their US peers of 1965 and 1976 on flexed arm hang scores, while both boys and girls outperformed their US counterparts on 1-minute speed sit-ups (partly because of differences in test structure). On the other hand, in the standing broad jump (where there is less scope for variations in the technique of testing), the 1979 CAHPER values are relatively close to the US national data reported by AAHPER in 1976 and other results obtained on Czechoslovakian students [Seliger and Bartunek, 1976]. After scaling results to allow for the increase of running distance, the CAHPER samples showed an improvement of sprint performance from 1964 to 1979; however, perhaps because of better track conditions, the US children out-performed their Canadian peers on this test.

Performance scores from the Trois Rivières longitudinal study are illustrated in figure 2. Scores for the control students were relatively similar to those for the national samples, but the figure also illustrates the very large increase of score which occurred in the experimental group, in association with an increase of physical education instruction.

English Canadian students have been studied cross-sectionally by Corroll et al. [1969] in Manitoba, and longitudinally by Ellis et al. [1975] in Saskatoon. The Manitoba data of 1969 were obtained for the most part on students who already had some familiarity with the 1964 CAHPER tests, and perhaps for this reason the scores tended to match the 1979 rather than the 1964 national data. In Saskatoon, students carried out the performance tests on an annual basis, and possibly because of practice

Table 6. National physical performance test data for 1964[1], 1979[2], and 1981[3] with Saskatoon data for 1970/1975[4]:

Age years	Speed sit ups (n)			Standing broad jump cm			Shuttle run s		Flexed arm hang s			45.9 m run, s	50 m run, s	274 m run, 6 s	800/1,600/ 2,400 m run, s
	1964	1970/1975	1979 1981	1964	1970/1975[4]	1979	1964	1979	1964	1970/1975	1979	1964	1979	1964	1979

Boys

Age															
6	–	–	19.5	–	–	112	–	15.8	–	–	15.8	–	11.7	–	340
7	19.8	–	24.2	112	–	120	14.1	15.2	22.9	–	18.9	10.2	11.0	85.8	320
8	23.5	–	29.6 27.8	122	–	129	13.5	14.5	28.4	–	25.0	9.7	10.3	80.5	291
9	26.4	–	33.4	132	–	136	13.0	13.9	31.4	–	30.8	9.2	9.9	76.7	275
10	27.3	37.8	35.8 35.3	137	164	146	12.8	13.5	32.8	30.6	32.9	8.9	9.5	73.8	575
11	30.2	39.4	37.9	147	169	154	12.5	13.1	34.9	36.1	36.0	8.6	9.2	71.9	555
12	29.7	41.9	37.9 38.9	152	178	163	12.3	12.6	37.5	46.4	37.3	8.5	8.9	69.2	548
13	32.5	43.3	41.2	163	187	175	12.0	12.3	41.0	50.9	42.9	8.1	8.5	66.0	802
14	32.9	45.1	41.3	178	195	188	11.6	11.8	47.1	56.1	48.0	7.7	8.1	62.5	766
15	33.4	47.8	42.3	188	209	196	11.3	11.6	49.6	65.7	50.8	7.3	7.8	60.0	744
16	34.9	48.9	42.3 38.6	201	218	210	10.9	11.3	53.2	64.5	55.4	7.1	7.5	57.8	722
17	34.1	–	42.5	208	–	216	10.8	11.2	52.6	–	54.5	7.0	7.4	56.2	710

Girls

Age															
6	–		19.3	–		107	–	16.1	–		12.9	–	12.0	–	364
7	17.2		24.0	109		115	14.8	15.7	18.4		14.9	10.7	11.4	87.6	341
8	18.8		29.2 26.4	117		125	14.1	15.0	17.8		18.1	10.0	10.6	81.7	322
9	20.2		31.4	124		132	13.8	14.4	19.3		22.1	9.7	10.1	79.1	311
10	22.4		33.8	132		139	13.3	14.0	21.5		22.4	9.2	9.9	76.4	664
11	24.8		34.6 32.7	140		149	13.0	13.5	21.1		23.6	8.9	9.5	74.1	642
12	23.1		35.7	142		156	13.0	13.2	18.7		22.9	8.8	9.2	72.5	641
13	22.7		36.4 34.0	147		162	12.8	13.0	17.6		23.0	8.7	8.8	71.9	983
14	20.1		36.0	150		165	12.7	12.8	16.4		23.7	8.7	8.7	73.1	959
15	22.2		34.7	155		165	12.6	12.9	16.5		23.5	8.5	8.7	72.4	973
16	22.5		35.2 31.8	157		169	12.5	12.7	15.8		24.8	8.5	8.6	72.0	946
17	19.9		34.7	152		172	12.6	12.7	15.8		27.0	8.5	8.6	72.9	941

[1] Hayden and Yuhasz [1966].
[2] Gauthier [1980].
[3] Canada Fitness Survey [1983].
[4] Ellis et al. [1975].

effects they obtained substantially higher scores than even the 1979 national sample with respect to flexed arm hang, speed sit-ups and standing long-jump.

Strength and Flexibility

The Canada Fitness Survey [1983] provides some national data for grip strength and flexibility, the latter measured by a sit and reach test (with a score of 25.0 cm corresponding to the touching of the floor with the out-stretched fingers).

The handgrip forces recorded by the Canadian students in 1981 (table 7) were some 5% higher than values obtained in Toronto [Shephard et al., 1969] and Edmonton [Howell et al., 1966] some 15 years earlier. They also appear higher than the values reported by Seliger and Bartunek [1975] from Czechoslovakia, although the comparison could here be vitiated by differences in the design of handgrip dynamometer (the Canadian measurements were made with the Stoelting instrument, which has a force plate individually adjustable for grip size).

The flexibility data are of a similar order to figures reported from various regional studies in the US [Shephard, 1986].

Discussion

Nutrition and Growth

The data discussed above document the existence of substantial regional differences of growth and development across Canada, with evidence of superimposed secular trends both nationally and within individual regions. However, it is less certain how far regional differences reflect ethnic groupings, and how far they can be attributed to differences of nutrition and other environmental factors.

A national nutritional survey [Nutrition Canada, 1973] indicated that the food intake of Canadian children was generally adequate. The estimated median energy intake was indeed generally higher than recommended by various expert groups; some special populations such as Inuit who were obtaining much of their food from hunting also had a remarkably high protein intake. Because of the difficulty in growing fresh vegetables in Canada for most of the year, serum levels of vitamin C were suboptimal both in 15–20% of the national sample and also in Indian bands living near major urban centers, while the vitamin C status of the more remote Inuit and Indian children was even less satisfactory. Vitamin A, vitamin D and calcium intakes were also below recommended levels in Inuit and Indian children, although there was no evidence that these deficiencies were sufficient to cause clinical abnormalities.

Table 7. Handgrip force (N) and flexibility (sit and reach test – board at 25.0 cm)

Completed age years	Handgrip force			Flexibility 1981[1]
	1981[1]	1968[2]	1966[3]	
Boys				
7–9	147	–	133	27.9
10–12	218	208	204	26.5
13–14	326	–	293	26.2
15–19	469	–	–	29.8
Girls				
7–9	132	–	131	31.8
10–12	192	181	187	30.9
13–14	268	–	248	32.8
15–19	294	–	–	34.1

[1] Canada Fitness Survey [1983].
[2] Shephard et al. [1969].
[3] Howell et al. [1966].

A possible influence of minor dietary deficiencies upon the growth and development of Canadian children can be inferred from differences of size between relatively prosperous Southern Ontario Indians and their poorer northern peers [Partington and Roberts, 1969]; in an earlier era, there were also substantial size differences between upper- and lower-class children in the Ottawa area [Hopkins, 1947]. The origins of the secular trend are complex [Shephard, 1982]. Nevertheless, a further pointer in the direction of significant nutritional imbalance is the overall secular trend to an increase of stature within Canada as economic conditions have improved. Secular trends to greater body size have been particularly rapid among French Canadian children. Francophones historically combined a small body size with a below average income and the responsibility of raising a large number of children per family. However, French Canadians have recently gained improved economic conditions, while reducing average family size below the English Canadian average [Léger and Lambert, 1985; Shephard, 1986].

Overnutrition and Inadequate Physical Activity

In the majority of Canadian children, overnutrition and a suboptimal level of activity pose a larger threat to development than malnutrition; in some, the early lesions of atherosclerosis are already established [Jaffé et al., 1971] soon after birth.

Unfortunately, there is little available information on the optimum percentage of body fat for either children or adolescents. One indication of overnutrition in city dwellers is the fact that skinfold readings are substantially lower in the Inuit children, without any apparent adverse influence upon their health. Again, the skinfold readings of the Igloolik population have increased as they have moved from an active, hunting lifestyle to a more sedentary pattern of existence. A further indicator of overnutrition is the transition from a relatively normal distribution of body fatness at the age of 6–7 years to a skewed distribution at the age of 12 years (fig. 1); such data seem to imply the emergence of an obese subpopulation. Some authors such as Amor [1978] have arbitrarily suggested that body fat should average 14% of body mass in adolescent males, and 18% in females, although in the Trois Rivières longitudinal study the breakpoint in the probit distribution curve occurs somewhat higher (at about 16% fat in the boys and 22% in the girls).

The need for a greater level of physical activity on the part of the average Canadian child is substantiated by the very high working capacity of traditional Inuit children, and the partial loss of this advantage with acculturation to a sedentary North American lifestyle. Confirmation of the need for enhanced endurance activity is seen in the accelerated physical development of Trois Rivières children who were enrolled in a daily programme of required physical education, and the equal advantage shown by English Canadian samples such as those of Cunningham and Eynon [1973] who were participating in endurance sport programmes.

Those interested in the promotion of fitness have been eager to attribute recent secular trends in working capacity and physical performance test scores to the more general promotion of physical activity on the part of the Canadian government. It is probable that federal and provincial motivational programmes have made some contribution in this regard, particularly with regard to the increase in PWC_{170} scores (where allowance is already made for effects of the parallel trend to an increase of body size). However, in the physical performance tests, a major part of the observed secular gains reflects a trend to increase of body size, and a further complicating factor has been the unavoidable influence of increased familiarity with the test items.

Canadian Children in a World Perspective

There are relatively few countries for which well-selected and representative population data are available. Canada is quite unusual with respect to the scope, scale and frequency of such representative sampling.

Some of the regional Canadian samples show interesting peculiarities, and the early Inuit data offer a possible insight into the development of physical performance which might be realized if other populations of children were to adopt a really active lifestyle. However, such special populations form no more than a small part of the Canadian mosaic. Not surprisingly, the overall national scores for growth, development and physical performance are generally comparable with available figures for other nations at a comparable stage of economic development. As in other parts of the industrialized world, development could be enhanced by an increase of physical activity and some decrease of total energy intake.

Conclusions and Recommendations

While differences of growth and development among the indigenous and founding peoples of Canada have traditionally been attributed to genetic factors, there is a need to review this dogma. Recent evidence points to a substantial and possibly an overriding influence of socioeconomic factors, including dietary practices. As overall economic conditions within Canada have improved and regional disparities of living conditions have disappeared, the trend has been to the emergence of children who are both larger for their age and of more uniform size from one ethnic group to another.

There remains a need to establish a clear optimal level of nutrition for the growing child. However, in Canada as in many other western nations, overnutrition currently seems a more important problem than undernutrition. Working capacity and physical performance test scores also fall short of potential, and programmes of enhanced physical activity should thus be encouraged among urbanised schoolchildren.

References

AAHPER: Youth Fitness Test Manual. Washington, American Association for Health, Physical Education & Recreation, 1958.
AAHPER: Youth Fitness Test Manual, revised ed. Washington, American Association for Health, Physical Education & Recreation, 1965.
AAHPER: Youth Fitness Test Manual, revised ed. Washington, American Association for Health, Physical Education, Recreation & Dance, 1976.
Alderman RB: Age and sex differences in PWC_{170} of Canadian school children. Res Q 1969;40:1–5.
Amor AF: A survey of physical fitness in the British Army; in Allen C (ed): Proceedings of the First RSG4 Fitness Symposium, with Special Reference to Military Forces. Downsview, Defence and Civil Institute of Environmental Medicine, 1978.
Bailey DA, Ross WD, Mirwald RL, Weese C: Size dissociation of maximal aerobic power during growth in boys; in Borms J, Hebbelinck M (eds): Pediatric Work Physiology. Medicine and Sport, Vol. 11. Basel, Karger, 1978, pp 140–151.

Canada Fitness Survey: Fitness and Lifestyle in Canada. Ottawa, Canadian Fitness and Lifestyle Research Institute, 1983.

Corroll VA, Nick G, LaPage R: Manitoba Physical and Motor Fitness Performance Manual. Winnipeg, Provincial Department of Youth & Education, 1969.

Cumming GR: Current Levels of Fitness. Can Med Assoc J 1967;96:868–877.

Cumming GR: Correlation of physical performance with laboratory measures of fitness; in Shephard RJ (ed): Frontiers of Fitness. Springfield, Thomas, 1971.

Cunningham DA, Eynon RB: The working capacity of young competitive swimmers 10–16 years of age. Med Sci Sports 1973;5:227–231.

Demirjian A, Bailey DA, de Pena J, Auger F, Jenicek M: Somatic growth of Canadian children of various ethnic origins. Can J Publ Health 1976;67:209–216.

Demirjian A, Jenicek J, Dubuc MB: Les normes staturo-pondérales de l'enfant urbain canadien-français d'âge scolaire. Can J Publ Health 1972;63:14–30.

Ellis JD, Carron AV, Bailey DA: Physical performance in boys from 10 through 16 years. Hum Biol 1975;47:263–281.

Gauthier R: Le manuel d'instruction du test d'efficience physique II de l'ACSEPR. Ottawa, Canadian Association for Health, Physical Education & Recreation, 1980.

Gauthier R, Massicotte D, Hermiston R, MacNab R: Comparaison entre 1968 et 1983 de la capacité physique de travail de jeunes Canadiens âgés de 7 à 17 ans. CAHPER J 1983;50:2–7.

Hayden FJ, Yuhasz MS: The CAHPER Fitness performance test manual for boys and girls 6 to 17 years of age. Ottawa, Canadian Association for Health, Physical Education & Recreation, 1966.

Hopkins JW: Height and weight of Ottawa elementary schoolchildren of two socio-economic strata. Hum Biol 1947;19:68–82.

Howell ML, Loiselle DL, Lucas WG: Strength of Edmonton schoolchildren. Edmonton, Fitness Research Unit, University of Alberta, 1966.

Howell ML, MacNab RBJ: The physical work capacity of Canadian children: 7–17 years. Ottawa, Canadian Association for Health, Physical Education & Recreation, 1968.

Jaffé D, Hartroft S, Manning M, Eleta G: Coronary arteries in newborn children. Intimal variations in longitudinal sections and the relationships to clinical and experimental data. Acta paediatr Scand 1971;(suppl 219):3–28.

Jeliffe DB: The Assessment of Nutritional Status of the Community. Geneva, World Health Organization, 1966.

Jenicek M, Demirjian A: Triceps and subscapular skinfold thickness in French-Canadian school-age children in Montreal. Am J Clin Nutr 1972;25:576–581.

Larivière G, Lavallée H, Shephard RJ: Correlations between field tests of performance and laboratory measures of fitness. Acta Paediatr Belg 1974;(suppl 28):19–28.

Léger L, Lambert J: Poids et taille des Québécois de 6 à 17 ans en 1981. Variations régionales, sexuelles et séculaires. Can J Publ Health 1985;76:388–397.

Mazess RB, Mather WE: Bone mineral content in Canadian Eskimos. Hum Biol 1975;47:45–63.

National Center for Health Statistics Growth Charts: Rockville, National Center for Health Statistics, 1976.

Nutrition Canada: Nutrition, a National Priority. Ottawa, Queen's Printer, 1973.

Partington MW, Roberts N: The heights and weights of Indian and Eskimo schoolchildren on James Bay and Hudson Bay. Can Med Assoc J 1969;100:502–509.

Rode A, Shephard RJ: Growth, development and fitness of the Canadian eskimo. Med Sci Sports 1973;5:161–169.

Rode A, Shephard RJ: Growth, development and acculturation – a ten year comparison of Canadian Inuit children. Hum Biol 1984;56:217–230.

Seliger V, Bartunek Z: Mean values of various indices of physical fitness in the investigation of Czechoslovak population age 12–55 years. Prague, CSTV, 1976.

Shephard RJ: Human Physiological Work Capacity. London, Cambridge University Press, 1978a.

Shephard RJ: Fitness, obesity and health: in Allen C (ed): Proc 1st RSG4 Physical Fitness Symp with Special Reference to Military Forces. Downsview, Defence and Civil Institute of Environmental Medicine, 1978b.

Shephard RJ: Physical Activity and Growth. Chicago, Year Book, 1982a.

Shephard RJ: Physiology and Biochemistry of Exercise. New York, Praeger, 1982b.

Shephard RJ: Fitness of a Nation. Lessons from the Canada Fitness Survey. Basel, Karger, 1986.

Shephard RJ, Allen C, Bar-Or O, Davies CTM, Degré S, Hedman R, Ishii K, Kaneko M, LaCour JR, DiPrampero PE, Seliger V: The working capacity of Toronto schoolchildren. Can Med Assoc J 1969;100:560–566, 705–714.

Shephard RJ, Goodman J, Rode A, Schaeffer O: Snowmobile use and decrease of stature among the Inuit. Arctic Med Res 1985;38:32–36.

Shephard RJ, Lavallée H, Larivière G, Rajic M, Brisson GR, Beaucage C, Jéquier J-C, LaBarre R: La capacité physique des enfants canadiens: une comparaison entre les enfants canadiens-français, canadiens-anglais et esquimaux. I. Consommation maximale d'oxygène et débit cardiaque. Union Méd 1974;103:1767–1777.

Skrobak-Kaczynski J, Lewin T: Secular changes in Lapps of Northern Finland, in Shephard RJ, Itoh S (eds): Circumpolar Health. Toronto, University of Toronto Press, 1976.

Stenett RG, Cram DM: Cross-sectional height and weight norms for a representative sample of urban, school-aged Ontario children. Can J Pub Health 1968;60:465–470.

Weiner JS, Lourie JA: Practical Human Biology. London, Academic Press, 1981.

Welch JP, Winsor EJ, MacKintosh SM: The distribution of height and weight and the influence of socio-economic factors in a sample of Eastern Canadian urban schoolchildren. Can J Publ Health 1971;62:373–380.

Prof. Roy J. Shephard, Director, School of Physical and Health Education, 320 Huron Street, University of Toronto, Toronto, Ont. M5S 1A1 (Canada)

Relationship of Dietary Intake to Work Output and Physical Performance in Czechoslovak Adolescents Adapted to Various Work Loads

Jana Pařízková, Jan Heller

Research Institute for Physical Education, Charles University, Prague, Czechoslovakia

Introduction

Living organisms react to various stimuli concerning energy input and output (for instance, dietary intake and work loads of various sorts) in differing ways before and after reaching maturity. Animal experiments started at a very early period in life indicate both immediate and delayed effects from manipulations of diet and exercise; these are not always of the same character or extent as those seen later in life, especially during adulthood [1, 2]. Experience with human subjects leads to similar conclusions. This applies, inter alia, to the adaptive consequences of exercise and athletic activities, which are commonly associated with specific changes of dietary intake.

There are marked differences in the morphological, functional and biochemical characteristics of subjects adapted to sports which develop predominantly cardiorespiratory efficiency and endurance, or skill, and/or muscular strength. Equally, dietary intakes vary considerably between athletes from various sport disciplines.

During recent years, the preparation of champion athletes has started quite early, certainly during adolescence, and in some disciplines much sooner (for instance, gymnastics, tennis, diving and skiing). This situation offers an interesting model for studying the impact of early adaptation to various work loads, including the consequences for dietary intake.

Subjects and Methods

Young athletes in contrasting sport disciplines (gymnasts and skiers of both sexes) were followed and compared with other groups such as divers, swimmers and hockey players. All subjects were enrolled in special training centres for youth, and had undertaken regular training in their disciplines for at least 2 years. Their physical performances were tested by both special tests characteristic of their discipline, and by evaluation of aerobic power.

Somatic development was compared to national standards [3] and the standards for body mass index (BMI = body mass kg/height m) of Rolland-Cachera and Sempé [4]. Body composition (the absolute and relative amounts of depot fat and lean body mass) was ascertained from skinfolds, using regression equations appropriate to the local population [1].

The dietary intake was monitored by the inventory method for at least 1 week. The results were analyzed using a special computer programme based on an analysis of local foodstuffs. The energy output was evaulated as suggested by WHO [5] (the various physical activities of each day were rated as multiples of basal metabolic rate as derived from body mass (BMR; WHO [5]) and were timed over each day). In addition, sport-testers evaluated the heart rate, with estimation of energy output from the relationship between heart rate and oxygen uptake. Energy output was thus assessed by indirect calorimetry, individual relationships being established for each subject [6, 7].

Results

Table 1 details the somatic and body composition characteristics in male (GM1) and female adolescent gymnasts (GF), followed during their preparatory training. Table 2 shows the absolute and relative intake of energy, proteins, fats and carbohydrates (related to total and lean body mass). Table 3 shows the intake of minerals and vitamins in gymnasts.

The energy output over this period was estimated as recommended by the WHO [5]; the mean calculated energy output in group GM1 averaged 14,500 kJ · day^{-1}, and in GF was 10,010 kJ · day^{-1}.

The energy output (GM2) was measured more exactly, along with dietary intake, in another group of boy gymnasts during a mountain training camp, where gymnastics (1.5 h · day^{-1}) was supplemented by ski training (5 h · day^{-1}). In this case sport-testers registered the pulse rate throughout the first day, and during the training period on subsequent days. The time spent on various activities (sleep, normal daily activities) and rest) was assessed, and the heart rate estimates of energy expenditure from day I were used to calculate total energy outputs. Under such conditions, group GM2 had a greater energy output than was normally found during gymnastic training.

Tables 4–6 provide the somatic and body composition characteristics, as well as the intake of energy, proteins, fats, carbohydrates, minerals and vitamins for group GM2, while the mean energy outputs from the 1st to the 6th days are given in figure 1. The average value was 18,277 ± 1,160 kJ · day^{-1}, which corresponds roughly with the average energy intake (table 5).

The somatic and body composition characteristics along with the intake of energy, proteins, fats, carbohydrates, minerals and vitamins for

Table 1. Somatic measures and body composition in gymnasts: mean, SD and variance of data

	Boys (n = 4)			Girls (n = 11)		
	$\bar{x}$	SD	$V\bar{x}$	$\bar{x}$	SD	$V\bar{x}$
Age, years	14.11	0.28	1.9	12.8	1.08	8.4
Height, cm	150.4	2.7	1.8	144.4	5.2	3.6
Body mass, kg	39.65	3.47	8.7	34.3	3.9	11.4
BMI	17.52	1.16	6.6	16.5	0.6	3.6
Fat, %	6.7	1.7	25.3	3.8	2.0	53.8
LBM, kg	36.9	3.1	8.4	32.9	3.6	11.0

BMI = Body mass index = body mass kg/ height m^2.

Table 2. The intake of energy, protein, fat and carbohydrate per day in gymnasts: mean, SD and variance of data

	Boys			Girls		
	$\bar{x}$	SD	$V\bar{x}$	$\bar{x}$	SD	$V\bar{x}$
Energy						
kJ	11,082	1,781	16.1	8,157	1,182	14.5
kcal	2,650	426	16.1	1,948	282	14.5
kJ · kg BM^{-1}	279	47	16.8	238	14	5.8
kJ · kg LBM^{-1}	300	51	17.0	247	15	6.0
Proteins						
g	75.0	7.6	10.1	74.1	6.1	8.2
g · kg BM^{-1}	1.9	0.4	21.1	2.1	0.4	18.5
g · kg LBM^{-1}	2.0	0.4	19.7	2.2	0.3	13.4
Fats						
g	119	21	17.4	100	17.5	17.4
g · kg BM^{-1}	3	0.6	20	2.9	0.6	20.0
g · kg LBM^{-1}	3	0.5	15.5	3	0.4	13.0
Carbohydrates						
g	328	58	17.7	193	32	16.0
g · kg BM^{-1}	8.2	1.8	21.8	5.6	1.5	25.0
g · kg LBM^{-1}	8.9	1.7	19.1	5.8	1.5	25.0

BM = Body mass; LBM = lean body mass;

Table 3. Mineral and vitamin intake per day in gymnasts: mean, SD and variance of data

	Boys			Girls		
	$\bar{x}$	SD	$V\bar{x}$	$\bar{x}$	SD	$V\bar{x}$
Minerals						
Ca, mg	594	148	24.9	679.9	129.1	19.0
Fe, mg	11.0	1.6	14.9	11.7	1.4	11.8
Vitamins						
A, µg	791	348	44.0	1,259	503	39.9
B_1, mg	1.02	0.05	4.9	1.4	0.2	14.3
B_2, mg	1.15	0.15	13.0	1.2	0.1	8.3
PP, mg	15.1	2.1	13.9	15.3	2.7	17.6
C, mg	42.5	6.0	14.1	60.9	33.7	55.3

Table 4. Somatic development and body composition in 6 gymnasts (training camp) (GM2): mean, SD and variance of data

	$\bar{x}$	SD	$V\bar{x}$
Age, years	15.57	1.7	10.9
Height, cm	157.8	6.3	4.0
Body mass, kg	48.6	9.3	19.1
BMI	19.4	2.2	12.4
Fat, %	9.0	1.6	17.7
LBM, kg	44.1	9.0	20.4

Table 5. Intake of energy, proteins, fats and carbohydrates per day in gymnasts (training camp): mean, SD and variance of data

	$\bar{x}$	SD	$V\bar{x}$
Energy			
kJ	18,145	1,142	6.3
kcal	4,338	273	6.3
$kJ \cdot kg\, BM^{-1}$	384	65	17
$kJ \cdot kg\, LBM^{-1}$	421	66	15
Proteins			
g	132.2	8.3	6.3
$g \cdot kg\, BM^{-1}$	2.8	0.5	16.8
$g \cdot kg\, LBM^{-1}$	3.0	0.5	15
Fats			
g	175	12.3	7.0
$g \cdot kg\, BM^{-1}$	3.7	0.7	19.5
$g \cdot kg\, LBM^{-1}$	3.9	0.6	16.0
Carbohydrates			
g	547.0	30.0	5.5
$g \cdot kg\, BM^{-1}$	11.7	2.3	20.0
$g \cdot kg\, LBM^{-1}$	12.8	2.4	19.0

Table 6. Mineral and vitamin intake per day in gymnasts (training camp): mean, SD and variance of data

	x̄	SD	Vx̄
Minerals			
Ca, mg	920.1	100.7	13.9
Fe, mg	20.6	1.5	7.2
Vitamins			
A, µg	992.7	94.1	9.5
B_1, mg	2.0	0.1	0.5
B_2, mg	2.1	0.3	14.3
PP, mg	33.5	6.5	19.4
C, mg	108.2	13.2	12.2

Table 7. Somatic measures and body composition in skiers: mean, SD and variance of data

	Boys (n = 4)			Girls (n = 6)		
	x̄	SD	Vx̄	x̄	SD	Vx̄
Age, years	16.62	1.40	8.4	15.7	1.21	7.71
Height, cm	180.7	3.13	1.7	164.1	5.6	3.4
Body mass, kg	69.3	3.1	4.9	49.3	7.2	14.6
BMI	20.5	0.54	2.6	18.2	1.6	8.7
Fat, %	5.2	1.81	34.8	10.8	2.3	21.5
LBM, kg	63.5	3.24	5.1	43.4	6.0	13.8

BMI = Body mass index = body mass kg/height m^2; LBM = lean body mass.

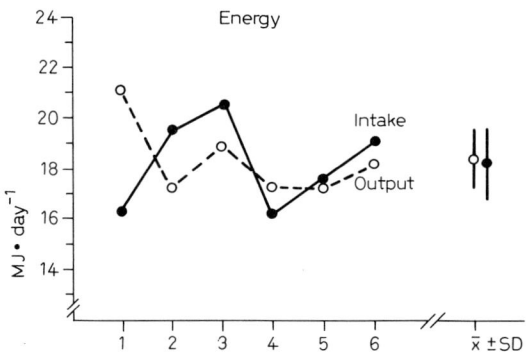

Fig. 1. Energy intake and output in male gymnasts (group GM2) during 6 days of training in the mountains.

Table 8. Intake of energy, proteins, fats and carbohydrates per day in skiers: mean, SD and variance of data

	Boys			Girls		
	$\bar{x}$	SD	$V\bar{x}$	$\bar{x}$	SD	$V\bar{x}$
Energy						
kJ	22,772	2,906	12.7	17,641	3,361	19.0
kcal	5,455	694	12.7	4,218	840	19.1
$kJ \cdot kg\ BM^{-1}$	338.9	32.4	9.6	358.3	42.4	11.8
$kJ \cdot kg\ LBM^{-1}$	357.8	29.1	8.1	405.4	39.3	9.7
Proteins						
g	154.6	15.6	10.1	116.1	21.2	18.3
$g \cdot kg\ BM^{-1}$	2.2	0.2	11.8	2.3	0.3	13.0
$g \cdot kg\ LBM^{-1}$	2.4	0.1	4.1	2.6	0.2	7.6
Fats						
g	215.4	27.7	12.8	168.0	50.8	30.2
$g \cdot kg\ BM^{-1}$	3.2	0.3	9.3	3.4	0.7	20.6
$g \cdot kg\ LBM^{-1}$	3.4	0.3	8.8	3.8	0.7	18.4
Carbohydrates						
g	751.8	105.4	14.0	578.2	83.2	14.4
$g \cdot kg\ BM^{-1}$	11.2	1.2	10.7	11.8	1.4	11.8
$g \cdot kg\ LBM^{-1}$	11.8	1.0	8.4	13.4	1.1	8.2

BM = Body mass; LBM = lean body mass.

Table 9. Mineral and vitamin intake per day in skiers: mean, SD and variance of data

	Boys			Girls		
	$\bar{x}$	SD	$V\bar{x}$	$\bar{x}$	SD	$V\bar{x}$
Minerals						
Ca, mg	1,636.3	263.9	16.0	1,111.3	260.4	23.4
Fe, mg	25.3	2.6	10.2	18.5	3.4	18.3
Vitamins						
A, µg	3,153	374	11.8	2,124	556	26.1
B_1, mg	2.7	0.2	7.4	2.1	0.3	14.3
B_2, mg	2.9	0.2	6.8	2.7	0.3	11.1
PP, mg	29.5	0.9	3.0	26.9	2.6	9.7
C, mg	151.7	43.6	28.7	126.9	13.0	10.2

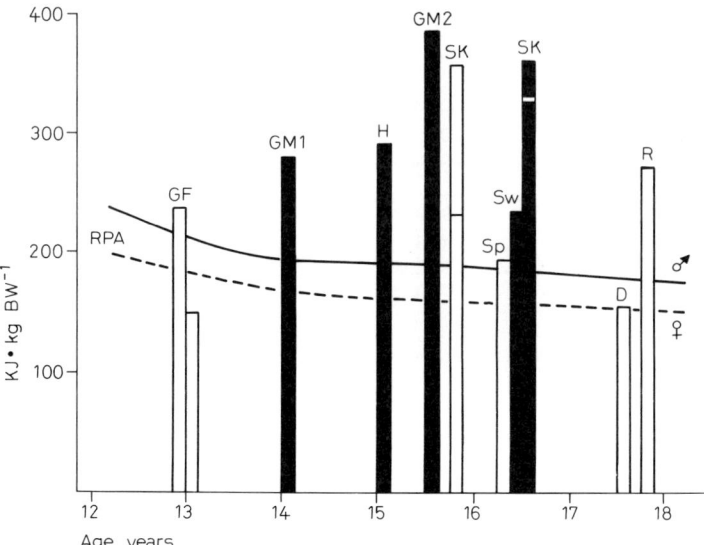

Fig. 2. The comparison of energy intake per kg body mass and day in groups of different adolescent athletes. White columns = females; black columns = males; GF, GM1 and GM2 = gymnasts; H = hockey players; SK = skiers; Sp = sprinters; D = divers; R = runners. Sk: lines within the columns = values ascertained during normal training in the centre; full length of the columns = values ascertained during training in the mountains. Full line = Recommended dietary allowance (RDA) for males; interrupted line = RDA for females of normal population [5].

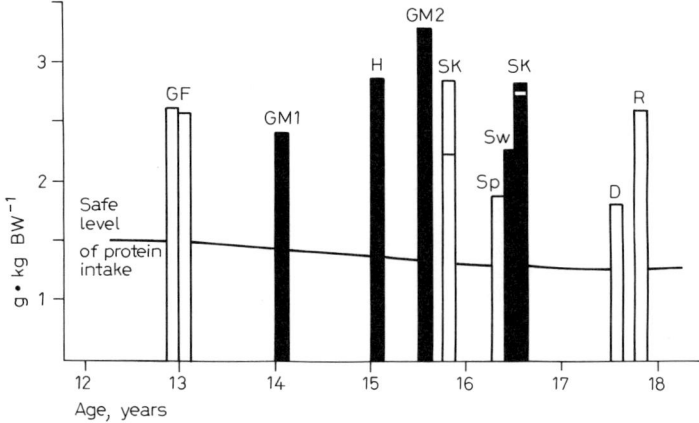

Fig. 3. The comparison of protein intake per kg body mass and day in groups of different adolescent athletes (for explanation see legend to fig. 2). Full line = Safe level of protein intake [5].

skiers of both sexes (SM and SF) are given in tables 7–9. The energy output was measured during 1 day in 6 female skiers during a training camp in the mountains ($18{,}975 \pm 2{,}446$ kJ · day^{-1}), The energy intake during the same period is given in table 8. In 2 male skiers the output was 26,477 kJ · day^{-1}, and the intake of energy 22,012 kJ · day^{-1}.

The energy and protein intake related to body mass (BM) in the above-mentioned subjects and in further groups of adolescent athletes may be compared with the RDA for a normal untrained population of the same age [5] (fig. 2, 3).

In skiers, the mechanical efficiency was $25.6\% \pm 1.8\%$ in group SF, and $29.6\% \pm 0.7\%$ in group SM. In group GM2 it was $26.0\% \pm 1.4\%$. In normal untrained subjects the mechanical efficiency under the same conditions is usually 20–21% [8].

Discussion

Differences in body size and body mass index (BMI) as compared to the national standards [3] appear mostly in the group of gymnasts; skiers differ less. Body height in gymnasts of both sexes corresponded to the 3rd percentile, BMI corresponded to the 25th percentile. This physique in gymnasts has been one of the predispositions to success in gymnastics, and may be explained mostly by selection based on hereditary factors. The body height of skiers of both sexes corresponded to the 50th percentile, which applies also to body height, proportionality and BMI in male skiers. The BMI corresponded to the 25th percentile in skiers of both sexes.

A significantly lower percentage of fat has been a regular phenomenon in nearly all trained athletes, including adolescents; this is apparent when comparing these subjects with data ascertained in the framework of the International Biological Programme (IBP) in subjects of a normal population in our country [9] (tables 1, 4, 7).

The data on energy intake were evaluated using both the absolute (per person) and relative values (per kg body mass and/or lean body mass). Marked differences appear among groups according to their sport specialization, and also during various periods of training. The same applies to the values of protein intake (fig. 2, 3), as well as those of fat and carbohydrates.

Energy balance was compared in the above-mentioned sport disciplines because of their differing character. When analyzing the differences between energy intake and output in girl gymnasts, a permanent deficit in energy intake of about 20% seemed to appear. But when comparing energy intake per kg body mass, it was slightly higher than in normal inactive girls of the same age (fig. 2). For comparison, values of energy intake are given for a different group of girl gymnasts, who tried temporarily to reduce weight and fatness; the energy intake was even lower at that period than in normal inactive girls. But such a reduction in the intake (i.e. $5{,}590 \pm 1{,}160$ kJ/day, roughly their BMR) lasted only a short time. Nevertheless, even when the deficit of energy intake seemed marked on a long-term basis, group GF continued to grow and train. Under conditions of continuous growth in height and proportionality in their particular growth channels [3], the health and physical performance of girl gymnasts was not disturbed [10].

In male gymnasts, this deficit (about 8%) was not so apparent (fig. 2) and diet seemed to be balanced over longer periods of time. The character of training in males requires, inter alia, great muscle strength; their physique is not so extreme as in girl gymnasts. In all instances the intake of proteins in gymnasts was even higher than is recommended for that period of growth (fig. 3).

The intake of energy in skiers (groups SM and SF) was one of the highest from all athletes ever measured in our laboratory. However, the intake of energy and other nutrients in gymnasts under special conditions of training in the mountains along with skiing (GM2) was the highest of all (fig. 2).

The evaluation of energy intake and output in the latter group of male gymnasts (GM2) showed satisfactory equilibration during 6 days. Energy intake increased only during the second day of training on skis, and resulted in an increase of energy output (fig. 1). When comparing the average values of energy intake and output during this period, the energy balance seemed adequate. The aerobic power in GM2 was also adequate (i.e. 60.2 ± 5.1 ml $O_2 \cdot kg^{-1} \cdot min^{-1}$) [6, 7, 8, 9].

When analyzing the protein intake data and comparing them with the RDA of the WHO [5] and/or those for the USA [11], an increased intake can be seen (fig. 3). The same applies to fats (tables 2, 5, 8). The intake of carbohydrates was too low in gymnasts, especially in girls (table 2), with the exception of group GM2.

The intake of minerals and vitamins was deficient in gymnasts (in some cases by more than by 50%), with the exception of vitamins A and B_1 and B_2 in girls. This was not true when the food intake was increased in group GM2 (table 6). In skiers, the intake was mostly excessive, along with an increased energy intake (fig. 2), which was nevertheless still approximately 16% lower than the estimated energy output. The only exception was the intake of Ca in girl skiers (SF), which was below the RDA, although it ought to be even higher because of growth and training. Up to now, no RDA have been available for growing athletes of different specializations.

During the period of our measurements, the body mass did not show substantial variation in all the groups followed; the same applied to their physical performance. This finding seems to indicate some adaptational adjustment as regards energy balance, and some increase of mechanical efficiency. The energy intake in some girl gymnasts (the interindividual variability was remarkable, as in normal untrained subjects) was only slightly higher than their estimated basal metabolic rate (BMR). Chronic deficits between energy intake and output have been described in other population groups, e.g. in East Java [12], or in Sherpa porters in the Himalayan mountains. Edmundson [12] assumes some 'compensatory mechanisms' that allow a much greater efficiency in those accustomed to low-energy intakes than in those with relatively high-energy intakes.

It is apparent, especially from the results in girl gymnasts, that the above-mentioned improvement in efficiency is already developing during the growth period, and when it remains within the adaptive limits of the organism, it may be tolerated without any major deterioration of the organism [13]. This applies especially when the composition of the ingested food corresponds in most respects with the recommended dietary allowances. But our data clearly indicate, that under conditions of reduced food intake, the preservation of an optimal composition may be very difficult; this may be a greater problem as regards optimal health and physical performance than the long-term slight reduction of energy intake to which the organism has adapted.

Previous observations on Tunisian children from poor families who had lower values of most somatic characteristics than their peers from affluent families and/or Czech children revealed better results for the step test and strength measurements and performance in some disciplines [1]. These data, along with others [14], indicate that under conditions of marginal malnutrition the functional capacity – especially cardiorespiratory efficiency and/or skill – does not necessarily deteriorate in spite of smaller body size. The same applies to the majority of gymnasts.

When the deficit is too extreme due to a lowered energy intake and/or an extreme work load, a reduction of functional capacity, stunting and health deterioration can occur. This was observed in some girl gymnasts, who finally had to interrupt their training because of some health problems, including amenorrhoea and retardation of skeletal maturation. It seems necessary to enforce changes in the rules and style of female gymnastics which would permit the highest achievements to those with a normal somatic development and dietary intake.

Skiing represents quite a different kind of sport, for which the development of endurance and thus an increased energy intake is necessary. The percentage of depot fat was reduced in the skiers and the aerobic power as characterized by the maximal oxygen intake was higher than in untrained subjects of the same age ($\mathring{V}_{O_2}$ max SF = 62.7 ± 3.3 ml · kg^{-1} · min^{-1}; SM = 66.5 ± 3.3 ml · kg^{-1} · min^{-1} [6, 7]. This is a favorable phenomenon not only from the point of view of physical performance, but also from the point of view of health. Repeated measurements of energy intake showed values in skiers as the highest ever measured in any athletes. This also appeared when gymnasts (GM2) were enrolled in skiing training. Åstrand and Rodahl [15] include skiing among the most energy-demanding sports. Work performed at a low environmental temperature obviously requires more energy than under other conditions. Comparison of energy intake and output in skiers corresponded better than in groups GM1 and GF; nevertheless, some deficits were occasionally found, under conditions of stable body mass, especially when energy needed for covering the expenses for growth was also considered.

However, the problem of possible 'compensatory mechanisms' [12] enabling greater efficiency of energy usage may be solved only when the above-mentioned deficits are verified by the use of more exact methods for the evaluation of energy output and the utilization of ingested foodstuffs, during long periods along with an assessment of contributions from internal sources of energy. This applies mostly to depot fat; therefore, not only body mass but also body composition has to be checked during and after a period of work.

Increased efficiency of energy utilization surely favours athletic performance, especially in activities demanding a very high energy output. The digestion of too much food costs energy, and moreover is not well tolerated over prolonged dynamic athletic performances [16]. This did not seem a problem in adolescent skiers during training; during the growth period the excess of food intake may be better tolerated than later on.

Further elucidation of ways of improving mechanical efficiency and of long-term adaptive processes originating early in life can help to achieve a higher level of power output and performance. Optimal work performance also has important economic implications under conditions of relatively low energy intake.

References

1 Pařízková J: Body Fat and Physical Fitness. The Hague, Martinus Nijhoff BV/Medical Division, 1977, pp 237–239, 251.
2 Pařízková J: Age dependent changes in dietary intake related to work output, physical fitness and body composition. Am J Clin Nutr 1989;89:962–967.
3 Prokopec M: The trend of somatic development of child population in Czechoslovakia during last thirty years. Čsl Hygiena 1986;31:541–557 (in Czech).
4 Roland-Cachera MF, Sempé M: Body Mass Index Charts. Paris, Institut National de la Santé et de la Récherche Médicale, 1985.

5   WHO: Energy and Protein Requirements. Report of a Joint FAO/WHO/UNU Expert Consultation. Tech Rep Ser 724. Geneva, World Health Organization, 1985.
6   Heller J, Novotný J, Bobvošová V, Bunc V, Pařízková J, Dlouhá R, Jirků L: Energy output during training preparation of cross-country skiers – juniors. Report on SPTR No 1-333-801/04, No. 12. Prague Research Institute for Physical Education, Charles University 1989 (in Czech).
7   Heller J, Novotný J, Bunc V, Pařízková J, Dlouhá R, Bobvošová V, Tůma L: Energy balance during training preparation of gymnasts – juniors. Report of SPTR No 1-333-801/04/06, No. 12. Prague, Research Institute for Physical Education, Charles University, 1989 (in Czech).
8   Bunc V, Heller J, Pařízková J, Šprynarová Š, Leso J: Changes of mechanical efficiency as a result of adaptation to an increased work load in different sports disciplines; in Dotson ChO, Humphrey GH (eds): Exercise Physiology, vol 2. New York, AMS Press, 1988, pp 133–143.
9   Seliger V, Bartůněk Z: Mean Values of Various Indices of Physical Fitness in the Investigation of Czechoslovak Population Aged 12–55 Years. Prague, Czechoslovak Association of Physical Culture, 1976.
10 Pařízková J: Body composition and nutrition in different types of athletes: in Taylor TG, Jenkins NK (eds): Proceedings of the XIII International Congress of Nutrition (Brighton 1985). London, Libbey, 1986, pp 309–311.
11 Recommended Dietary Allowances, ed 9. Washington, National Academy of Sciences, 1980, p 186.
12 Edmundson W: Individual variations in work output per unit energy intake in East Java. Ecol Food Nutr 1977;8:147–151.
13 Pařízková J: Growth, functional capacity and physical fitness in normal and malnourished children: in Bourne GH (ed): Nutrition in Health and Disease, Wld Rev Nutr. Basel, Karger, 1987, pp 1–44.
14 Spurr GB: Nutritional status and physical work capacity. Yearbook Phys Anthropol 1983;26:1–35.
15 Åstrand PO, Rodahl K: Textbook of Work Physiology, New York, McGraw-Hill, 1977, pp 665.
16 Brouns F, Saris WHM, Stroecken J, Beckers E, Thijssen R, Rehrer NJ, ten Hoor F: Eating, drinking and cycling. A controlled Tour de France Simulation study. II. Effect of diet manipulation; in Saris WHM (ed): Nutrition and Top Sport. Int J Sports Med 1989; (suppl I)10:S41–S48.

Jana Pařízková, MD, DSc, Research Institute for Physical Education,
Charles University (VÚT UK), Ujezd 450, 11807 Prague 1 (Czechoslovakia)

Conclusions

Roy J. Shephard

School of Physical and Health Education, University of Toronto, Ont., Canada

The present volume presents new information on the growth of body build and physical work capacity in children who are at a nutritional handicap, whether from a low socioeconomic status or as a consequence of intensive physical activity (child laborers and residents of athletic training camps).

It should immediately be emphasized that the studies of the third-world poor reported here have been limited largely to agricultural populations and children of the 'working poor'. It is discouraging to recognise that a parent earning the pittance of $36/month is still receiving four times the minimum wage in a country such as Brazil, where many people live in considerable luxury. Although the patterns of growth and development observed in underprivileged third-world environments are appreciably different from those previously described in affluent European and North American samples, it is also important to recognise that a much greater impact of malnutrition would have been seen if studies had been extended to those children whose parents earned only the minimum wage, to the 'street children' of the developing world, and to children in countries plagued by famine and civil war.

Nevertheless, the present monograph documents a considerable stunting of growth in both stature and body mass relative to US standards, even in communities where income is regarded as moderate, and no gross deficiencies of either total energy intake or protein availability have been demonstrated. In general, the limitation of growth in body mass seems roughly proportional to the limitation of linear growth; in other words the problem is one of stunting rather than a frank wasting of skeletal muscles. Nevertheless, physical performance remains poor even after making a simple linear adjustment of data for differences of height and age relative to standards for developed countries. Mathematicians may object to linear height adjustments, since much recent research suggests that muscle force and aerobic power develop as the third power of stature [Shephard et al., 1980]; however, the difference between a linear and a power function relationship is small when the difference of height is only a few centimeters, and investigators from third-world countries should not be faulted for making simpler analyses within the competence of their computing facilities.

The limitation of physical performance in the poor child may have an immediate negative impact upon academic education, because self-image is impaired by comparisons of athletic ability with students from more affluent areas. It is also difficult to be sure that the poor scores reported for both physical performance tests and psychological assessments have not been limited by the immediate effects of hunger, poor motivation, a noncompetitive culture, and the lack of stimulation in the home environment.

The nutritionally deprived child tends to grow into a small adult, often with a low level of acquired intelligence as measured by standard tests (although such tests are admittedly geared to people of high socioeconomic status). The poor attainment on intelligence-measuring instruments may reflect poor motivation, and school time that has been lost through the need to help the parents in child care, primitive agriculture, or as a child laborer, rather than a direct effect of malnutrition upon the growth and development of the brain. Nevertheless, the end-result of poor intellectual development is a person who is handicapped in adult life. The high wage jobs of the technical world are closed to this individual, and any employment must be sought as a physical laborer. Because of small adult size, there may also be a limited ability to undertake physical labor, with a further negative influence upon both employability and the capacity to grow food for the next generation. There is thus a strong likelihood that the vicious cycle of poverty will be passed from one generation to the next. This unfortunate tendency is reinforced by the fact that in many third-world countries, simple manual occupations provide substantially less than a living wage. A combination of physical limitations and intellectual deprivation prevent the poor family from breaking out of such a bondage.

Some nutritionists have commented on the ability of the body to 'adapt' to shortages of food. Certainly, a small, malnourished person has a lower basal metabolic rate, and expends less energy in displacing body mass over a fixed distance. The latter phenomenon is an inevitable consequence of basic Newtonian laws, but it can hardly be considered a physiological adaptation. Extreme malnutrition may also bring about a reduction of basal metabolic rate per unit of body mass, and a starving community tends to sink into a torpor, where very little voluntary activity is undertaken. The third-world person is thus small, and eats less food than a European or North American; but such an individual fails to achieve their potential, and should not be considered as 'small but healthy'. Observations of this sort should certainly not be used to justify a world economic system where the majority of children fail to reach possible limits of growth, being held to a small, undernourished physique.

Special sports schools inevitably impose a high energy demand on students, whether they are found in the developed or the third world. We have left unanswered the difficult ethical questions of the desirability of ruthless, all-out international competition and the concentration of a child's athletic experience in a highly focussed school from an early age. But if such institutions are to continue (as seems likely), then there is plainly a need for upward revision of recommended dietary allowances for the students concerned. The detemination of an appropriate dietary allowance for the student athlete nevertheless remains a puzzle. The normality of growth curves can hardly provide the answer, because the student athletes are highly selected, and often begin their sporting careers with unusual characteristics of height and body mass. Moreover, puberty is often advanced or retarded in such individuals, so that a misleading impression of the rate of growth is obtained if longitudinal data are compared with observations on sedentary students of similar age. Recourse is at present made to nitrogen balance studies and vitamin saturation tests, but such procedures are quire difficult to carry out, and they are also subject to many pitfalls. Further consideration of the dietary needs of the very active child is thus required.

Finally, we should not neglect the dietary needs of the average student in our developed societies. Often, the problem for such individuals is overnutrition rather than undernutrition. It is worth stressing that recommended dietary allowances incorporate a substantial safety margin (all suggested intakes are 2 SDs above the average need). Thus, if a child receives the full RDA, but yet is at the lower end of the physical activity spectrum, considerable overnutrition is likely to occur. This has adverse consequences for future health; in particular, the accumulation of fat increases the risk of ischemic heart disease as an adult. Some studies of North Americans have shown that atherosclerotic lesions of the major blood vessels are already quite advanced by the third decade of life, and such adverse changes can only be avoided by treating atherosclerosis as a disease of childhood. In Western Europe and North America, the average child needs to seek a more appropriate balance of energy intake and output. While the risk of fat accumulation could be reduced by either an increase of physical activity or a reduced consumption of food, there are strong arguments to favour the first of these options [Shephard, 1982], and there is an urgent need to devise patterns of physical activity that will appeal to city-dwelling children on a life-long basis.

References

Shephard RJ: Physiology and Biochemistry of Exercise. New York, Praeger Publishing, 1982.

Subject Index

Anaerobic power 49–54
Algeria and growth 61–78
Anaerobic threshold 106, 107
Anthropometry
  body composition 5, 43–45
  physical performance 110–113
Argentine Republic, physical fitness 80–98
Athletes
  body mass 19–21
  dietary intake 157–166
  height 19–21

Basal metabolic rate 13–16, 24–26, 165
Blood hemoglobin, Colombian children 52–54, 57, 58
Body
  composition 7, 8, 24, 25
    Mexico 119–130
  mass 148
    Algerians 65–68
    Canadians 140, 141
    Chinese 20, 21
    index 7, 163
Brazil, nutrition 109–116

Calcium insufficiency 29
Canada, fitness 133–153
Cardiac morpnology, athletes 23
Cardiopulmonary function, athletes 22–24
Caste levels, nutrition 36–39
Child laborers 36, 41
Childhood stress, social class 36
Children, health 1
China 19–31
Close sports specialization 103, 106
Colombia, nutrition 41–58
Computer-assisted tomography 8
Cuba, physical capacity 99–107
Czechoslovakia, workloads 156–166

Dietary intake, nutrition 12–18
Diseases of civilization 2

Electrical conductivity, body mass determination 8
Energy
  cost of exercise 26
  expenditure 13–16, 157–166
  intakes 27, 28
  requirements 16, 68
    exercise 24, 26
Environment, Algeria 75, 76
Exercise, nutrition 19–27

Fat mass 121, 123, 126–130
Fat-free mass 121, 123–130
Fifty meter dash 74, 75, 90, 91
Fitness assessment 4–12
Flexibility 150, 151

Gross domestic products, Algeria 61, 62
Growth 168
  environmental conditions 1–18
  exercise 19–21, 26
  India 34
  velocities, Colombia 45, 47, 48
Gymnast
  dietary intake 158–160, 164
  fitness 21

Handgrip force 71, 127, 150, 151
Heart rate 101
Height
  Canadians 138–141
  Chinese 20, 21
  Colombians 45, 46
Hydrostatic weighing 8

Inadequate physical activity 151, 152
India, social epidemiology of nutrition 33–39

Subject Index

Infant mortality rate, Algeria 62
Iron deficiency 29, 30

Lean body mass 45, 46, 50, 51, 56
Literacy levels, nutrition 37, 38

Malnutrition 109
  fitness 11, 18
Mass/height ratios 67
Maximal oxygen debt 106, 107
Maximal oxygen intake 10, 20, 43, 49, 106, 107, 144–146
Mean grip strength 127
Measurement techniques 134
Median body mass 5–8
Mexico, body composition 119–130
Mineral requirements 27, 29, 159–161, 164
Morphofunctional values 86–89
Morphological development 5–8
Motor performance, nutrition 114
Multiple correlation analyses, nutrition 37
Muscle strength 11, 12

National Plan of Physical Fitness Evaluation 80
Neutron activation techniques 8
Nitrogen balance 27, 28
Nutrition
  Algeria 61–78
  Brazil 109–116
  Canada 150, 151
  China 27–31
  dietary intake 12–18
  education, Algeria 78
  growth 55–58

Overnutrition 151–152

Physical condition, aerobic power 49–52, 54
Physical fitness 8–12
  Argentine Republic 80–97
  Canada 133–153
  Cuba 99–107
  socioeconomic strata 80–97
Physical performance 146–150, 168, 169
Physical work capacity 10, 11, 69–71
  Canada 141, 143, 144

Cuba 99, 102
Population 2
Power output 100, 101
Protein
  energy malnutrition 109, 110
  intakes 28
  requirements 12, 17, 18, 26, 27

Recommended dietary allowances 3, 12, 13, 16–18, 26–31, 114
Rhythmic sports 103–105
Run 85, 93, 95, 96
Runners, fitness 22

Sexual maturation, Colombian children 48
Sit-ups 72, 73
Skiers, dietary intake 160–166
Skinfold measurements 8, 122, 124, 141, 142, 152
Soccer players, fitness 20, 23
Social epidemiology, nutrition 33–39
Socioeconomic strata, physical fitness 80–97
Specific dynamic action, exercise 26
Standing height 64
Standing long jump 72, 73
Step test 11
Strength, flexibility 150
Subcutaneous fat, Canadians 141

Team games 103–105
Total body conductivity measurements 8
Trace elements 29, 30

Undernutrition
  Brazil 119–130
  Colombia 44
  India 33–37

Ventricular hypertropy, training 23
Vital capacity 67–69
Vitamin(s) 30
  A 30
  B_1 30
  B_2 30
  requirements 27, 29, 159–161, 164

Youth, health 1

Zinc deficiency 29, 30